Hydrologic Effects of a Changing Forest Landscape

Committee on Hydrologic Impacts of Forest Management

Water Science and Technology Board

Division on Earth and Life Studies

NATIONAL RESEARCH COUNCIL
OF THE NATIONAL ACADEMIES

THE NATIONAL ACADEMIES PRESS
Washington, D.C.
www.nap.edu

THE NATIONAL ACADEMIES PRESS 500 Fifth Street, N.W. Washington, DC 20001

NOTICE: The project that is the subject of this report was approved by the Governing Board of the National Research Council, whose members are drawn from the councils of the National Academy of Sciences, the National Academy of Engineering, and the Institute of Medicine. The members of the committee responsible for the report were chosen for their special competences and with regard for appropriate balance.

Support for this project was provided by Department of the Interior Awards No. INTR-7397. Any opinions, findings, conclusions, or recommendations expressed in this publication are those of the author(s) and do not necessarily reflect the views of the organizations or agencies that provided support for the project.

International Standard Book Number-13: 978-0-309-12108-8 (Book)
International Standard Book Number-10: 0-309-12108-6 (Book)
International Standard Book Number-13: 978-0-309-12109-5 (PDF)
International Standard Book Number-10: 0-309-12109-4 (PDF)
Library of Congress Control Number: 2008940394

Hydrologic Effects of a Changing Forest Landscape is available from the National Academies Press, 500 Fifth Street, N.W., Lockbox 285, Washington, DC 20055; (800) 624-6242 or (202) 334-3313 (in the Washington metropolitan area); Internet, http://www.nap.edu.

Copyright 2008 by the National Academy of Sciences. All rights reserved.

Printed in the United States of America

THE NATIONAL ACADEMIES
Advisers to the Nation on Science, Engineering, and Medicine

The **National Academy of Sciences** is a private, nonprofit, self-perpetuating society of distinguished scholars engaged in scientific and engineering research, dedicated to the furtherance of science and technology and to their use for the general welfare. Upon the authority of the charter granted to it by the Congress in 1863, the Academy has a mandate that requires it to advise the federal government on scientific and technical matters. Dr. Ralph J. Cicerone is president of the National Academy of Sciences.

The **National Academy of Engineering** was established in 1964, under the charter of the National Academy of Sciences, as a parallel organization of outstanding engineers. It is autonomous in its administration and in the selection of its members, sharing with the National Academy of Sciences the responsibility for advising the federal government. The National Academy of Engineering also sponsors engineering programs aimed at meeting national needs, encourages education and research, and recognizes the superior achievement of engineers. Dr. Charles M. Vest is president of the National Academy of Engineering.

The **Institute of Medicine** was established in 1970 by the National Academy of Sciences to secure the services of eminent members of appropriate professions in the examination of policy matters pertaining to the health of the public. The Institute acts under the responsibility given to the National Academy of Sciences by its congressional charter to be an adviser to the federal government and, upon its own initiative, to identify issues of medical care, research, and education. Dr. Harvey V. Fineberg is president of the Institute of Medicine.

The **National Research Council** was organized by the National Academy of Sciences in 1916 to associate the broad community of science and technology with the Academy's purposes of furthering knowledge and advising the federal government. Functioning in accordance with general policies determined by the Academy, the Council has become the principal operating agency of both the National Academy of Sciences and the National Academy of Engineering in providing services to the government, the public, and the scientific and engineering communities. The Council is administered jointly by both Academies and the Institute of Medicine. Dr. Ralph J. Cicerone and Dr. Charles M. Vest are chair and vice-chair, respectively, of the National Research Council.

www.national-academies.org

COMMITTEE ON HYDROLOGIC IMPACTS OF FOREST MANAGEMENT

PAUL K. BARTEN, *Chair*, University of Massachusetts, Amherst
JULIA A. JONES, *Vice-Chair*, Oregon State University, Corvallis
GAIL L. ACHTERMAN, Oregon State University, Corvallis
KENNETH N. BROOKS, University of Minnesota, St. Paul
IRENA F. CREED, The University of Western Ontario, Canada
PETER F. FFOLLIOTT, University of Arizona, Tucson
ANNE HAIRSTON-STRANG, Maryland Department of Natural Resources, Annapolis
MICHAEL C. KAVANAUGH, Malcolm Pirnie, Inc., Emeryville, California
LEE MACDONALD, Colorado State University, Fort Collins
RONALD C. SMITH, Tuskegee University, Tuskegee, Alabama
DANIEL B. TINKER, University of Wyoming, Laramie
SUZANNE B. WALKER, Azimuth Forest Services, Shelbyville, Texas
BEVERLEY C. WEMPLE, University of Vermont, Burlington
GEORGE H. WEYERHAEUSER, JR., Weyerhaeuser Company, Washington

National Research Council Staff

LAUREN E. ALEXANDER, Study Director
ELLEN A. DE GUZMAN, Research Associate
JULIE VANO, Consultant

WATER SCIENCE AND TECHNOLOGY BOARD

CLAIRE WELTY, *Chair,* University of Maryland, Baltimore County
JOAN G. EHRENFELD, Rutgers University, New Brunswick, New Jersey
SIMON GONZALEZ, National Autonomous University of Mexico, Mexico
CHARLES N. HAAS, Drexel University, Philadelphia, Pennsylvania
THEODORE L. HULLAR, Cornell University, Ithaca, New York
KIMBERLY L. JONES, Howard University, Washington, D.C.
G. TRACY MEHAN III, The Cadmus Group, Inc., Arlington, Virginia
JAMES K. MITCHELL, Virginia Polytechnic Institute and State University, Blacksburg
DAVID H. MOREAU, University of North Carolina, Chapel Hill
JAMES M. HUGHES, Emory University, Atlanta, Georgia
LEONARD SHABMAN, Resources for the Future, Washington, D.C.
DONALD I. SIEGEL, Syracuse University, New York
SOROOSH SOROOSHIAN, University of California, Irvine
HAME M. WATT, Independent Consultant, Washington, D.C.
JAMES L. WESCOAT, JR., University of Illinois at Urbana-Champaign
GARRET P. WESTERHOFF, Malcolm Pirnie, Inc., White Plains, New York

Staff

STEPHEN D. PARKER, Director
LAUREN E. ALEXANDER, Senior Staff Officer
LAURA J. EHLERS, Senior Staff Officer
JEFFREY W. JACOBS, Senior Staff Officer
STEPHANIE E. JOHNSON, Senior Staff Officer
WILLIAM S. LOGAN, Senior Staff Officer
M. JEANNE AQUILINO, Financial and Administrative Associate
ANITA A. HALL, Senior Program Associate
ELLEN A. DE GUZMAN, Research Associate
DOROTHY K. WEIR, Research Associate
MICHAEL J. STOEVER, Project Assistant

Preface

In 1976, a group of Forest Service scientists[1] published a seminal volume on forests and water that evaluated the effects of forest management on floods, sedimentation, and water supply. It was one of the first comprehensive attempts to link upstream forest management with downstream water management and supply. For many years, this report served as a critical reference for forest hydrology scientists and managers. Much has changed since 1976. Thirty years ago, no one would have imagined that clear-cutting on public lands in the Pacific Northwest would come to a screeching halt, that farmers would give up water for endangered fish and birds, or that climate change would produce quantifiable changes in forest structure, species, and water supplies. Today, however, these phenomena shape the management of forests and water. Such developments have sharpened public awareness and heightened tensions between water users and water sources. It is time to enumerate these changing factors and assess the science of forest hydrology in light of these dynamic circumstances.

The forest hydrology literature is full of articles that have attempted to synthesize, retrospectively, the hydrologic effects of forest management. This literature reflects sharply different views of the magnitude and significance of those effects. To date, there have been no examples in the published literature in which a group of scientists, managers, and practitioners came together and reached consensus on what is known and what needs to be known about the hydrologic effects of forest disturbance and management. This National Research Council (NRC) report, *Hydrologic Effects of a Changing Forest Landscape*, does just that. This report combines forest and water management perspectives, but the committee strove to exclude value judgments on forest management practices, watershed management, and water augmentation. The NRC committee structure was essential to realizing this major contribution to the forest and water communities because it provided a unique setting for convening experts to reach consensus on these important and timely issues.

The members of this NRC committee include scientists, engineers, practitioners, and policy experts from across North America, each with unique perspectives on and knowledge of forests and water. The committee embraced this rare opportunity to integrate diverse expertise and synthesize collective knowledge in a form that advances science and water resource management. The committee hopes that this report is valuable to the next generation of

---

[1] Anderson, H.W., M.D. Hoover, and K.G. Reinhart. 1976. Forests and water: Effects of forest management on floods, sedimentation, and water supply. USDA Forest Service, Pacific Southwest Forest and Range Experiment Station, General Technical Report PSW-18/1976. Berkeley, Calif., 115 pp.

scientists, land and water managers, and citizens who strive to sustain water resources from forests in the coming years.

The committee's work was made possible by the essential support of the staff of the Water Science and Technology Board (WSTB). Dr. Lauren Alexander, WSTB senior staff officer and study director, played a critical leadership role on behalf of the committee to focus and moderate our discussions and clarify the report's message at all stages of committee deliberations, report preparation, and completion. The committee is extremely grateful for her skilled management of ideas, her lucid writing, and the long hours she dedicated to the completion of this report. Thanks also go to Julie Vano, an NRC science and technology policy graduate fellow and now a doctoral student at the University of Washington for her valuable contributions.

Ellen de Guzman, research associate, sets the standard for efficiency, grace, and field-expedient problem solving in relation to meeting and field trip arrangements, research support, assistance in responding to review, and report publication. Stephen Parker, WSTB director, actively participated in several meetings and lent experience and advice to the chair, committee, and staff. The vice chair of this committee, Dr. Julia Jones of Oregon State University, deserves special mention and praise for her tireless commitment to this report, its clear message, and its completion. Dr. Jones was a central, guiding, and uniting force for the committee, and the committee and staff are most appreciative of her intellectual contributions and commitment to building consensus. Above all, the committee expresses its gratitude for her leadership on this project.

Speakers and presenters shared valuable information, insights, and perspectives on the committee's work during our information gathering meetings in Nevada, Colorado, Oregon, Georgia, and Washington, D.C. We thank Paul Adams, Oregon State University; Maryanne Bach, Bureau of Reclamation; David Bayles, Pacific Rivers Council; Jarylyn Beek, Bureau of Reclamation; Peter Bisson, U.S. Forest Service; Mike Cloughesy, Oregon Forest Resources Institute; Karl Cordova, Rocky Mountain National Park; Terry Cundy, Potlatch Corporation; Clayton Derby, Platte River Cooperative Agreement; Kelly Elder, U.S. Forest Service; Dallas Emch, Willamette National Forest and Central Cascades Adaptive Management Partnership; Megan Finnessy, McKenzie Watershed Council; Cheryl Friesen, Willamette National Forest and Central Cascades Adaptive Management Partnership; Gordon Grant, U.S. Forest Service; Stan Gregory, Oregon State University; Deborah Hayes, U.S. Forest Service; Polly Hays, U.S. Forest Service; Michelle Isenberg, BASF Corporation; Rhett Jackson, University of Georgia; Linda Joyce, U.S. Forest Service; Calvin Joyner, U.S. Forest Service; Randy Karstaedt, U.S. Forest Service; Dave Kretzing, McKenzie River Ranger District; John Lawson, Bureau of Reclamation; Kara Lamb, Bureau of Reclamation; Dan Levish, Bureau of Reclamation; Ted Lorensen, Oregon Department of Forestry; Chris Jansen Lute, Bureau of Reclamation; Deborah Martin, U.S. Geological Survey; Dale

McCullough, Columbia River Inter-Tribal Fish Commission; Karl Morgenstern, Eugene Water & Electric Board; Ted Oldenburg, Hoopa Valley Tribal Forestry, California; Fred Ore, Bureau of Reclamation; Nancy Parker, Bureau of Reclamation; Maryanne Reiter, Weyerhaeuser Company; Matt Rea, U.S. Army Corps of Engineers; Arne Skaugset III, Oregon State University; Thomas Spies, U.S. Forest Service; Brian Staab, U.S. Forest Service; John Stednick, Colorado State University; Randall Stone, Massachusetts Department of Conservation and Recreation; Fred Swanson, U.S. Forest Service; Rick Swanson, U.S. Forest Service; Brad Taylor, Eugene Water & Electric Board; Albert Todd, U.S. Forest Service; Charles Troendle, U.S. Forest Service (Retired); and Eric Wilkinson, Northern Colorado Water Conservancy District.

This report has been reviewed in draft form by individuals chosen for their diverse perspectives and technical expertise, in accordance with procedures approved by the NRC's Report Review Committee. The purpose of this independent review is to provide candid and critical comments that will assist the institution in making its published report as sound as possible and to ensure that the report meets institutional standards for objectivity, evidence, and responsiveness to the study charge. The review comments and draft manuscript remain confidential to protect the integrity of the deliberative process. We wish to thank the following individuals for their review of this report: Robert Beschta, University of Oregon; Terry Cundy, Potlatch Corporation; Thomas Dunne, University of California, Santa Barbara; Rhett Jackson, University of Georgia; J. B. Ruhl, Florida State University; Thomas D. Kyker-Snowman, Massachusetts Department of Conservation and Recreation; John D. Stednick, Colorado State University; and David A. Woolhiser, consultant.

Although the reviewers listed above have provided many constructive comments and suggestions, they were not asked to endorse the conclusions and recommendations nor did they see the final draft of the report before its release. The review of this report was overseen by Margaret B. Davis, Emeritus, University of Minnesota. Appointed by the National Research Council, she was responsible for making certain that an independent examination of this report was carried out in accordance with institutional procedures and that all review comments were carefully considered. Responsibility for the final content of this report rests entirely with the authoring committee and the institution.

Table of Contents

Summary 1

1 FORESTS, WATER, AND PEOPLE 13
   Forests 14
   Movement of Water Through Forests 15
   Forest Hydrology 17
   The NRC Study of Hydrologic Effects of Forest Management ... 20

2 FORESTS AND WATER MANAGEMENT IN
     THE UNITED STATES 23
   Forests and Water in the United States 23
   Managing Forests and Water 27
   Emerging Issues for Forests and Water 33
   Summary 41

3 FOREST DISTURBANCE AND MANAGEMENT EFFECTS ON
     HYDROLOGY 45
   Forest Hydrology Science 45
   Modifiers of Forest Hydrology 47
   Hydrologic Responses: General Principles 50
   Hydrologic Responses Within Forests 51
   Changes in Watershed Outputs 53
   Managing Forests for Water 73

4 FROM PRINCIPLES TO PREDICTION: RESEARCH NEEDS FOR
     FOREST HYDROLOGY AND MANAGEMENT 75
   Spatial Research Needs 75
   Temporal Research Needs 78
   Social Research Needs 83
   Cumulative Watershed Effects 85
   Summary 87

5 RECOMMENDATIONS FOR FORESTS AND WATER IN THE
     TWENTY-FIRST CENTURY 95
   Recommendations for Forest Hydrology Scientists 95
   Recommendations for Managers 101
   Recommendations for Citizens and Communities 104

Moving Forward: Forest Hydrology Science and Management in the
Twenty-First Century.. 109
REFERENCES 113

APPENDIXES 143

A Institutional Governance and Regulations of Forests and
Water................…... 145
B Committee Biographical Information.. ..163

Summary

The forests of the United States cover about one-third of the country's land area and are managed for a number of purposes—timber harvesting, wilderness, habitat, and recreation—but arguably their most important output is water. Precipitation is cycled through forests and soil, and ultimately some is delivered as streamflow to receiving bodies of water. In this way, forests process nearly two-thirds of the freshwater supply in the United States.

Demand for water in the United States is increasing, and forest managers today are asked to provide higher quantities and qualities of water. Water supply managers question whether different land use management in forested headwaters can help meet downstream water quantity or quality demands. Meeting water supply needs is becoming more difficult because elevated water demand is occurring simultaneously with changes in climate, human population and development, land use, and ownership. How to manage forests and sustain water supplies will be a primary challenge in the twenty-first century.

The science of forest hydrology investigates the rates and pathways of water movement through forests. Forest hydrology researchers have amassed a comprehensive understanding of how water is connected to and moves through forests. A strong evidence base has emerged for understanding basic processes and principles of water movement through forests that can be used to predict the general directions and magnitudes of hydrologic effects of changes in forest cover, climate, and land use.

As the demand for water increases in the United States, water managers increasingly draw upon this strong scientific foundation and seek input from the forest hydrology community to identify ways to ensure reliable supplies of water. The U.S. Department of the Interior's Bureau of Reclamation is the largest wholesaler of water in the United States, providing water for more than 31 million people and 10 million acres of irrigated farmland. The U.S. Forest Service (USFS) manages 193 million acres of land for a continuing supply of timber, favorable conditions for streamflow, recreation, wilderness areas, and other objectives. These two agencies requested that the Water Science and Technology Board of the National Research Council (NRC) convene a committee to study and produce a report on the present understanding of forest hydrology, connections between forest management and attendant hydrologic effects, and directions for future research and management needs to sustain water resources from forested landscapes (see Statement of Task, Box S-1). In response, the NRC appointed the Committee on Hydrologic Effects of Forest Management, a group of 14 experts, to generate this report.

> **BOX S-1**
> **Statement of Task**
>
> This study will examine the effects of forest management on water quantity, quality, and timing. The report will reflect on the state of knowledge, relevant policy implications, and research needs that would advance understanding of connections among hydrology, science, and land management and policy in forested landscapes.
>
> 1. What is the state of knowledge of forest hydrology?
> 2. What are information and research needs regarding forest hydrology in forested lands?
>   - Topics could include sediment-related watershed processes; surface and groundwater hydrology; biological and ecological aspects; and extrapolation of small-scale study results to large-scale management practices.
>
> 3. What are the new issues that need to be addressed to ensure clean and plentiful water?
>   - Topics could include extreme weather events, climate change, fire, and invasive species.
>
> 4. How well are forest hydrologic impacts understood over short and long temporal scales and small and large spatial scales?

STATE OF FOREST HYDROLOGY SCIENCE

Forest hydrology is the study of water in forests: its distribution, storage, movement, and quality; hydrologic processes within forested areas; and the delivery of water from forested areas. Forest hydrology research uses field measurements, experiments, and modeling to characterize and predict hydrologic processes and their responses to natural disturbance and management of forests. It draws upon disciplinary knowledge from several branches of hydrological sciences, water resources engineering, and forestry to address primary questions of forests and water:

- What are the flowpaths and storage reservoirs of water in forests and forest watersheds?
- How do modifications of forest vegetation influence water flowpaths and storage?
- How do changes in forests affect water quantity and quality?

"Paired watershed" studies have been a primary empirical approach in forest hydrology. In paired watershed studies, two watersheds that are similar in size, initial land use or land cover, and other attributes are selected for study; both are monitored—one is then left as "control," and the other is "treated" (i.e., subject to manipulations such as forest cutting, road building, etc.). The measured changes in the relationship of streamflow and water quality between the treated and the control watersheds quantify the effects of forest treatment and regrowth. Most paired watershed studies in forest hydrology were begun in the 1940s, 1950s, and 1960s, but many of these studies were discontinued in the 1980s.

Paired watershed studies, process measurements, plot-scale studies, and hydrologic modeling are important elements of forest hydrology science. Study plots and paired watershed experiments generally range in size from less than a square meter to 1-2 km^2, and time scales for plot and process studies most commonly span only a few growing seasons. However, some USFS Experimental Forests and Ranges have conducted watershed studies spanning several decades or longer, particularly those designated as Long-Term Ecological Research (LTER) sites, funded by the National Science Foundation.

Forest Hydrology Processes and General Principles

Forest hydrology studies show that changes in forest structure and composition, and associated changes in forest soils and hillslopes, can alter the storage and flowpaths of water through soil and subsoil, which modifies water yield, peak flows, low flows, water chemistry, and water quality (see Figure S-1). The general principles of water movement in and through forests are understood with a high level of certainty (Table S-1).

Using Forest Hydrology Science to Inform Management Decisions

The current body of forest hydrology science supports forest and water management decisions in many ways. Forest hydrology science has led to a clear understanding of general principles (Figure S-1, Table S-1) of water movement through forests. These principles indicate the general magnitudes and directions of direct hydrologic responses to changes in forests over short time scales and in small areas. However, today's forest and water managers need forest hydrology science to predict or indicate the indirect and interacting hydrologic responses in forest landscapes that are changing over large areas or long time scales.

A pressing question for forest hydrologists is whether cutting trees in forested headwaters will augment water yield downstream for agricultural, municipal, or other uses while maintaining desired ecological attributes associated with forested landscapes. Although it is possible to increase water yield by harvesting timber, the increases in water yield from vegetation removal are often small and unsustainable, and timber harvest of areas sufficiently large to augment water yield can reduce water quality. The potential for increasing water yield from forest management is low, which reflects that increases are less likely in seasons when water demand is high and increases tend to be much smaller in drier years.

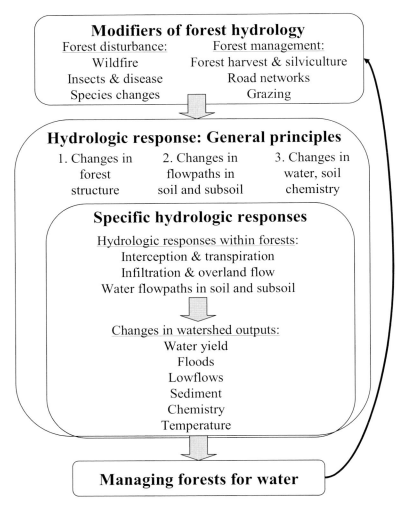

FIGURE S-1 Forest hydrology examines the flowpaths and storage of water in forests and how forest disturbance and management modify hydrologic responses. Hydrologic responses to changes in forests fall into three categories of general principles, as well as specific hydrologic responses, discussed in the text. The final section of this chapter evaluates the state of knowledge of forest hydrology and its implications for managing forests for water, including feedbacks to processes that modify forests.

TABLE S-1 General Principles of Forest Hydrology Describing the Direct Effects on Hydrologic Processes of Changes in Forest Structure, Changes in Water Flowpaths, and Application of Chemicals

	Principles of Hydrologic Response to Changes in Forest Structure
1	Partial or complete removal of the forest canopy decreases interception and increases net precipitation arriving at the soil surface
2	Partial or complete removal of the forest canopy reduces transpiration
3	Reductions in interception and transpiration increase soil moisture, water availability to plants, and water yield
4	Increased soil moisture and loss of root strength reduce slope stability
5	Increases in water yield after forest harvesting are transitory and decrease over time as forests regrow
6	When young forests with higher annual transpiration losses replace older forests with lower transpiration losses, this change results in reduced water yield as the new forest grows to maturity
	Changes in Water Flowpaths in Soils and Subsoils
7	Impervious surfaces (roads and trails) and altered hillslope contours (cutslopes and fillslopes) modify water flowpaths, increase overland flow, and deliver overland flow directly to stream channels
8	Impervious surfaces increase surface erosion.
9	Altered hillslope contours and modified water flowpaths along roads increase mass wasting
	Hydrologic Response to Application of Chemicals
10	Forest chemicals can adversely affect aquatic ecosystems especially if they are applied directly to water bodies or wet soils
11	Forest chemicals (fertilizers, herbicides, insecticides, fire retardants) affect water quality based on the type of chemical, its toxicity, rates of movement, and persistence in soil and water
12	Chronic applications of chemicals through atmospheric deposition of nitrogen and sulfur acidify forest soils, deplete soil nutrients, adversely affect forest health, and degrade water quality, with potentially toxic effects on aquatic organisms

NOTE: These general principles are not predictions, so qualifying adjectives such as "may," "usually," etc. are omitted. See Chapter 3 for factors that influence when, where, and to what extent these principles apply.

RESEARCH NEEDS IN FOREST HYDROLOGY

To meet the needs of the managers and users of forests and water, forest hydrology research has to move from principles to prediction. Predictions are needed to understand the indirect and interacting hydrologic responses to changes in forested landscapes associated with climate change, forest disturbances, forest species composition and structure, and land development and ownership, and how these changes will affect water quantity and quality downstream and over long time scales.

A Landscape Approach to Forest Hydrology

A landscape perspective on forest hydrology links scientific principles from plot, process, and small watershed scales with indirect and interacting hydrologic responses at larger spatial scales (i.e., within drainage basins and across large climatic and physiographic regions) in forest landscapes that are changing over long time scales. Within watersheds, forests are located in headwaters and downstream areas, on hillslopes and in riparian zones, and forests fulfill different water-related functions depending on their location. A key unresolved issue in forest hydrology is how to "scale up" findings from one part of a watershed to larger areas or to the entire watershed.

The temporal context for a landscape approach to forest hydrology involves expanding the temporal scale into the past to quantify the effects of antecedent forest management and disturbances and into the future to project and anticipate changes in land use and climate. For example, past forest harvest practices, road networks, fire suppression policies, grazing practices, and natural disturbances such as fire and wind have left legacies in forest structure and composition. These legacies affect hydrologic processes.

The research needs for a landscape approach to forest hydrology science involve studies that determine the following:

- How general principles developed in small, homogeneous watersheds can be used to improve predictions of hydrologic responses across large, heterogeneous watersheds and landscapes;
- How forests and forest management activities affect hydrologic processes, runoff, and water quality as a result of their position within a watershed;
- How local effects of roads can be scaled up to quantify the effects of road networks on water quantity and quality in larger watersheds and regions, particularly during large storms; and
- How long-term legacies of forest disturbance and forest management practices affect forests, water quantity, and water quality.

Forest Disturbance

Forests are dynamic ecosystems subject to both incremental and episodic disturbances that vary in frequency, severity, and extent. Probable hydrologic responses to fire, insects and disease can be inferred from the general principles of forest hydrology (Table S-1). However, compared to the extensive literature on hydrologic responses to forest management, relatively few studies have examined hydrologic responses to fire, insects, and disease in forests, especially at long time scales or in large watersheds.

The research needs for understanding hydrologic effects of forest disturbances involve studies that determine

- Effects of high- versus low-severity forest fires on water quantity, quality, and flooding, and how these effects vary over time and spatial scales; and

- Hydrologic responses to interacting and cumulative effects of forest disturbance (such as fire and insect outbreaks) and forest management (including thinning, salvage logging, roads, timber harvesting, and fire suppression).

Forest Management

Much of the forest hydrology literature focuses on the hydrologic effects of timber management practices and roads. Forest management practices evolve over time, resulting in new practices, such as thinning for fuel reduction, and best management practices (BMPs), such as managing wider riparian buffers for species protection. Moreover, recent increases in fire, insects, and disease in forests have spurred the adoption of forest management practices, such as thinning and salvage logging, whose effects on hydrology have received little study. The hydrologic effects of many of the new management practices and BMPs have not been studied, and dynamic forest conditions make it important to understand how contemporary practices influence water resources.

Research needs for understanding hydrologic responses to forest management involve:

- Studies that determine how contemporary forest management on public and private lands affects water quantity and quality and
- Improved forest hydrology models that reliably simulate the hydrologic and water quality responses of watersheds in varied forest conditions.

Cumulative Watershed Effects

One of the biggest threats to forests, and the water that derives from them, is the permanent conversion of forested land to residential, industrial, commercial, and infrastructure uses. Cumulative watershed effects (CWEs) include the hydrologic effects resulting from multiple land use activities over time within a watershed. Assessing CWEs requires an understanding of the physical, chemical, and biological processes that route water, sediment, nutrients, pollutants, and other materials from hillslopes and headwater streams to downstream areas. CWE research strives to establish cause-effect relationships among forests, water, and watersheds over large spatial and temporal scales.

Research needs for CWEs involve the following:

- A landscape-scale approach to relate downstream conditions to changes in forest conditions and land use in the contributing watershed; and
- Spatially explicit models that identify, connect, and aggregate changes due to forest disturbance and management over time in large watersheds.

Climate Change

Some effects of climate change on forests and water are already evident, and future climate changes are likely to have major effects on forest hydrology. Observed direct effects of climate warming on forests and hydrology include such as changes in the timing of snowmelt runoff and increases in wildfires. More research is needed to better predict indirect effects of climate change, including evaluations of how changes in forests and forest management influence hydrologic response.

The research needs related to the hydrologic effects of climate change include:

- Direct effects of climate change on hydrologic processes in forests and on water yield and water quality from forests;
- Indirect effects of climate change on forest structure and species composition and the consequences of these changes for water yield and water quality; and
- Indirect effects of climate change on forest disturbance, including wildfires, insects and diseases, and the consequences of these changes for water yield and water quality.

RECOMMENDATIONS TO SUSTAIN WATER RESOURCES FROM FORESTS

Scientists who study forest hydrology, forest and water managers, and citizens who use water can take actions to sustain water resources from forests. Each of these groups has important roles to play in applying the current understanding, exploring research gaps and information needs, and pursuing recommended actions (Table S-2).

Recommendations for Scientists

Scientists are poised to advance forest hydrology science to address critical water issues. New research approaches should be pursued in addition to maintaining and expanding existing data. In doing so, **scientists should:**

- Continue current small watershed experiments;
- Reestablish small watershed experiments where research has been discontinued;
- Centralize historical records from watershed studies in digital, well-documented, publicly accessible databases;
- Use the whole body of paired watershed data as a "meta-experiment" to

better understand and improve utility for managers of hydrologic responses to forest disturbance and management over large spatial and temporal scales and a range of forest types;

- Expand the capability for visualization and increase the prediction accuracy of hydrologic response in large watersheds through geographic information systems (GIS), remote sensing, sensor networks, and advanced models; and
- Work with economists and social scientists to improve and communicate understanding of the value of sustaining water resources from forests.

Recommendations for Managers

Managers of forests and water play critical roles in providing water resources from forests. Because forests, forest management, and the climatic and social contexts of forests are dynamic, BMPs must be updated continually through an adaptive management approach. Forestry BMPs can mitigate the negative consequences of forest management activities (roads, timber harvest, etc.), but their effectiveness can be highly site- and storm-specific or difficult to quantify. Forest and water managers are well positioned to use rigorous monitoring to assess the effectiveness of BMPs. In response to their assessments, managers can adapt management approaches and modify the current suite of BMPs to increase their effectiveness and test the results.

To assist the evolution of BMPs, managers should:

- Catalogue individual or agency BMP use, design, and goals at the national level and make this information available to the public;
- Monitor BMP activities for effectiveness, and coordinate analyses of monitoring data for use in an adaptive management framework; and
- Design adaptive management approaches for forested watersheds that coordinate management, research, monitoring, and modeling efforts.

Recommendations for Citizens

Cumulative watershed effects, changes in land ownership and management, changing population and development patterns, and water supply concerns have spurred activity to protect watersheds and water quality from the grass-roots, community level. New community-level watershed councils and forest groups are proactive in watershed-based restoration and management. Water researchers and policy makers have long recognized the benefits of organizing land and water management around watersheds and taking an integrated approach to watershed management. An integrated watershed management approach can help track the effects of various land uses on water supply and quality. Citizens and communities can influence forest and water management at the local, regional, or watershed level.

TABLE S-2 Current Understanding, Research Needs, and Recommendations for Sustaining Water Supplies from Forests

	Current Understanding	Information Gaps and Research Needs	Recommended Actions
Science	The body of forest hydrology science derives from almost 100 years of studies at small spatial and time scales Forest hydrology science has established general principles that are understood with a high degree of certainty describing direct hydrologic effects of forest management and disturbance Effects can be understood through changes in • Forest structure • Magnitudes, rates, and flowpaths • Erosion, nutrient cycling, and soil chemistry Reduced forest cover results in increased water yield that is • Generally short-lived • Greatest during times of water excess rather than water scarcity • Small or undetectable in water-scarce areas • May be associated with a decline in water quality	Hydrologic effects of past management, such as fire suppression, clear-cutting, roads Ways to quantify hydrologic responses at larger spatial and temporal scales Ways to scale up findings from small spatial and short time scales to larger spatial and longer time scales Use general principles to predict indirect hydrologic responses to changes in forest landscapes and interacting responses to forest management and disturbance	Enhance, maintain, and reestablish abandoned small watershed studies Combine existing data from the large body of small watershed studies and analyze them for large-scale trends as a meta-experiment Use new technologies, including sensor networks and remote sensing, to improve understanding of forest hydrology in changing landscapes Engage in adaptive management to help managers and community groups design monitoring strategies, develop and test models, and conduct studies relevant to management
Management	Forests in the United States are managed for a wide range of goals and objectives: timber harvesting,	Assessment of BMP effectiveness Principles and practices of adaptive management	Advance BMP evolution by rigorously assessing and developing new BMPs and

			measuring their effectiveness At the federal level, provide sustained support for adaptive management activities, enabling managers to partner with scientists to design and implement monitoring, develop and test models, and conduct studies relevant to management issues Increase role of agency technical expertise in watershed councils
	road networks and road construction, high-severity wildfires, and exurban sprawl modify forest hydrology Forest management practices are evolving in response to environmental change, social and economic forces, and technological developments BMPs are used to mitigate impacts on water resources from forest management activities		
Community	Integrated watershed management is a viable vehicle for both community groups and state and federal agencies to help manage water and forest resources at the community scale Citizens groups can influence local and integrated watershed management Community watershed groups benefit from state and federal agency technical expertise Existing laws can be used to strengthen the standing and influence of watershed councils New laws offer increased opportunities for community involvement	How watershed councils and their stakeholders view and utilize forest hydrology science and scientific expertise from federal agencies How industry-sponsored green certification and federal forest stewardship contracts affect water quantity and quality from forests	Use watershed councils to meet multiple goals of integrated watershed management at the community level Expand the number and influence of watershed councils. Engage in adaptive management with scientists and managers

Watershed councils and citizen groups should work within communities and with state and federal agencies to:

- Use watershed councils as vehicles to meet multiple goals of integrated watershed management at the community level; and
- Participate in watershed councils and help them grow in number and influence over watershed uses at the community level.

CLOSING

Forest hydrology science has produced a solid foundation of general principles that describe how water is connected to and moves through forests and how hydrologic processes respond to forest disturbance and forest management. The forest landscape is dynamic: it is continually changing in response to climate, natural disturbance, and forest management, as well as demographics and development patterns. Forest hydrology science and management are adapting as land use and ownership within forested watersheds become more heterogeneous, changes in climate and its effects are becoming more evident, and new technologies provide improved capability to predict and visualize cumulative watershed effects over larger spatial scales and longer periods of time. Building on the strong foundation of general principles of forest hydrology, new forest hydrology research can fill information gaps in the coming decades (Table S-2). Forests are essential for the sustainable provision of water to the nation. It is incumbent upon scientists, policy makers, land and water managers, and citizens to use the lessons of the past and apply emerging research, technology, and partnerships to protect and sustain water resources from forested landscapes.

1
Forests, Water, and People

The connections among forests, water, and people are strong: forests cycle water from precipitation through soil and ultimately deliver it as streamflow that is used to supply nearly two-thirds of the clean water in the United States. This connection between forests and water is not always tension-free. In fact, in many areas across the United States, water-related tensions are growing.

In one case—the North Platte River Basin in Colorado and the Rocky Mountain region—the tension is about headwater sources and how, if at all, manipulations of land uses in the forested headwaters produce changes in the water supply. Water scarcity contributes additional strain to this situation because water demand in the North Platte River Basin exceeds the allotted water supply from the river most of the time. The U.S. Bureau of Reclamation is responsible for ensuring adequate supplies of water to other water managers and suppliers, irrigators and agriculturalists, hydropower generators and users, and municipalities. The U.S. Forest Service (USFS) and other landowners, recreationalists who utilize headwater areas, and those with concerns about the environment and endangered species, are pressed to change their land use and management practices to increase the amount of water available downstream. These issues are not academic: both upstream and downstream stakeholders recognize the strength of the connections between forests and water and their access to how these connections affect water.

In another case on the other side of the Mississippi River, a similar tension is felt. In West Virginia, an extreme weather event dropped more than 6.5 inches (165 mm) of rain in a single storm in July 2001. The resulting floods caused extensive property damage and worse, death. The impacted downstream residents filed a lawsuit that claimed timber harvesting, among other headwater land uses, caused or contributed to the devastating flood damage, and the state-appointed Flood Protection Task Force concluded that forest harvesting operations may have affected flood flows, and the major flood risk was associated with logging roads and culvert designs. In both of these cases, as well as many others across the United States, the science of forest hydrology may provide valuable inputs in understanding and resolving these tensions.

The science of forest hydrology investigates the rates and pathways of water movement through forests. In most parts of the country, as in the North Platte Basin and West Virginia, forested headwater areas are a primary source of water supply. In the United States in the twentieth century, per capita water use increased from less than 10 to more than 75 gallons per day, and water demand per acre of forest increased by five- to twentyfold. Society's growing demand for clean water and healthy ecosystems, combined with tensions related to water supply or flooding risks, challenge forest hydrologists to predict how changes in a forest will affect the quantity and quality of water to help meet that demand. These challenges are becoming more acute as water demand increases simultaneously with changes in climate, land use, and other processes in forest systems. This report discusses these challenges and provides the scientific basis and context for addressing them using a suite of recommendations for the scientist, the forest or water manager, and the citizenry.

FORESTS

Forests account for 33 percent of all U.S. land area, covering about 750 million acres (300 million hectares) (Powell et al., 1993; Smith et al., 2004). Of this, 57 percent (430 million hectares) are privately owned, and the remainder is public forest. The federal government owns or manages land in all 50 states, with its largest holdings concentrated in 13 western states.

The forest products industry is an important element of the global economy, accounting for approximately $200 billion each year. In the United States, timber harvesting operations produce nearly 400 million cubic meters of wood annually. Forests also provide recreational opportunities and aesthetic values, carbon sequestration and mitigation of some air pollutants, and fish and wildlife habitat. Forest management plans and programs must address fire, drought, insect and diseases, habitat protection, wilderness areas, and recreation. All of these activities can have measurable influences on water supply and quality for municipalities, agriculture, and aquatic ecosystems from the channel to the watershed and landscape scales.

Forests are also efficient, low-maintenance, solar-powered living filters that provide high-quality water supplies that support aquatic ecosystems. Precipitation that comes as rain or snow in forested areas is cycled back to the atmosphere or drains through the soil to streams and aquifers, thereby producing much of the nation's water supply. In this way, forested areas provide water to 40 percent of all municipalities (Nulty, 2008) or about 180 million people in the United States (*http://www.fs.fed.us*).

The Forest Reserves Act (1891), the Organic Act (1897), and the Weeks Act (1911) first designated and established management of national forests. Since then, the social, economic, and political changes of the twentieth century, especially after World War II, increased the number, scope, and complexity of laws and regulations that guide the management of public and private forests. In addition to favorable conditions of flow and a continuing supply of timber, the USFS today must manage national forests for multiple objectives. These management responsibilities are sometimes supported and sometimes constrained by an increased understanding of forest- and water-related ecosystem services: natural filtration by vegetation and soils, provision of species habitat, groundwater and streamflow regulation, erosion control, and channel stabilization.

Forests that once provided high-quality runoff are becoming developed parcels that can adversely affect runoff patterns and water quality. Many of these ownership and use conversions occur through discrete, small parcels, such that land use change is hard to detect and has been easy to underestimate. Piecemeal changes in forest land use produce cumulative watershed effects that may be considerable and challenging to mitigate.

Climate change has potentially large but uncertain effects on forests and the water they process. Specific hydrologic effects of climate change on forests are complex and vary based on regional characteristics. The most important, widespread, and immediate effects of climate change are in the shift from snow to rain. In areas such the western United States that depend on snowpack for seasonal reservoir, the reduction of seasonal snow storage is expected to shift peak runoff earlier in the spring and reduce summer water availability to agriculture and cities. Climate change may increase favorable conditions for forest fires, outbreaks of insects and disease, and changes in forest structure and species composition, producing indirect hydrologic effects.

MOVEMENT OF WATER THROUGH FORESTS

A few basic principles form the foundation for the science of forest hydrology. Forest hydrologists use concepts of "balances" or "budgets" of water, energy, sediment, and nutrients, to understand how forests affect water quantity and quality. The degree to which the effects of forest management modify water quantity and quality over the long term has been the subject of forest hydrology studies for the past century. The resultant literature of forest hydrology is large, with consensus on many topics.

The water balance traces the transformation of precipitation (input) to runoff (output), which is of interest to the general public and water managers (Figure 1-1). The amount of precipitation is the dominant control on the amount of runoff. The timing and type of precipitation—rain, snow, or fog drip—also affect the amount and timing of runoff. A second major control on runoff is the transfer of water to the atmosphere by evaporation and transpiration from vegetation including trees (evapotranspiration, or ET), and a third control on runoff is the amount of water that can be infiltrated and stored (Figure 1-1). A third control on runoff is the amount of water that is stored or flows as groundwater (referred in this report as sub-surface flow), i.e., water that infiltrates into the soil surface; water that is stored in the soil profile, and water that moves laterally as groundwater flow (Figure 1-1). Although surface and groundwater hydrology are undoubtedly connected, forest hydrology and therefore, hydrologic effects of changes in forest cover, more strongly focus on surface flow, sub-surface flow within a few meters of the ground surface, infiltration, and overland flow.

The amount and timing of runoff are controlled in part by the water used by vegetation, which in turn depends on the amount of heat gained and lost by the system (energy balance; Figure 1-2). The energy budget influences air, soil, and

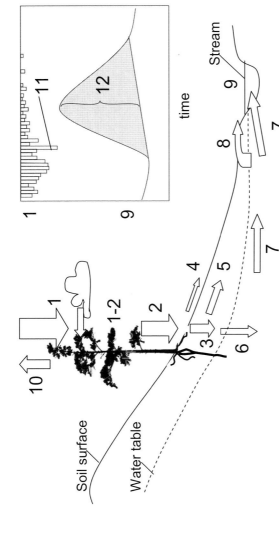

FIGURE 1-1 Elements of the water balance in a forest: 1 = precipitation (rain, snow, cloudwater deposition); 2 = net precipitation; 1- 2 = interception; 3 = infiltration; 4 = surface runoff, or infiltration excess (Horton) overland flow; 5 = subsurface flow, or lateral subsurface flow; 6 = groundwater recharge; 7 = groundwater flow; 8 = saturation excess overland flow; 9 = discharge or streamflow; 10 = evapotranspiration; 11 = precipitation intensity; 12 = peak flow or peak discharge. Although it is not shown, understory vegetation also contributes to these processes.

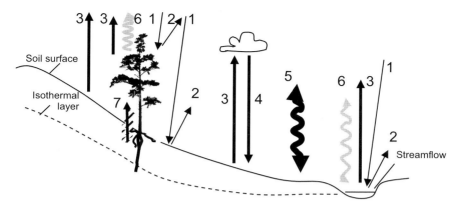

FIGURE 1-2 Elements of the energy balance in a forest. 1 = insolation (incoming shortwave radiation); 2 = reflection (of shortwave radiation due to albedo or reflectivity of vegetation, soil, and water surfaces); 3 = longwave radiation emitted by the Earth; 4 = longwave radiation reflected back to Earth from greenhouse gases including water vapor and CO_2; 1-2-3 + 4 = net radiation. Radiation inputs into the forest may be transformed into sensible heat (5), resulting in warming of the environment, latent heat (6, the energy consumed in evapotranspiration), or metabolic heat (7, the energy stored in biochemical reactions). Although it is not shown, understory vegetation also contributes to these processes.

water temperatures and drives key processes such as photosynthesis and transpiration. In snow-dominated systems, snowmelt is a primary hydrologic consideration, and energy exchange at the snowpack surface influences the rate and timing of runoff.

Water quality from forests depends on the flowpaths and the budgets of water, sediment, and nutrients within ecosystems (Figures 1-3 and 1-4). An understanding of these flowpaths and constituents of water quality is needed to predict forest water quantity, quality, and delivery processes in the majestic redwood and Douglas fir forests in the Pacific Northwest, the taiga forests in Alaska, the snow-dominated spruce and pine forests in the Rocky Mountains, or the broad-leafed, deciduous forests in the eastern United States.

FOREST HYDROLOGY

Forest hydrologists employ multiple approaches to study the pathways and fates of water, energy, sediment, and nutrients; these are called "process studies." Watershed studies examine (1) inputs and outputs of water, sediment, and nutrients, and (2) forest management activities and forest change. Modeling studies test process understanding and allow predictions.

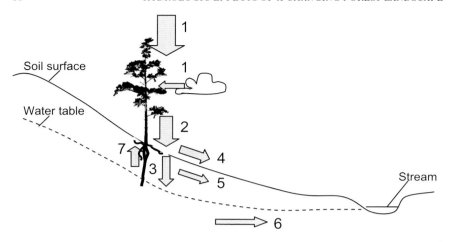

FIGURE 1-3 Pathways of the sediment and nutrient budgets in a forest: 1 = atmospheric deposition; 2 = net deposition; 1-2 = interception; 3 = immobilization in soil; 4 = surface erosion; 5 = shallow mass movements (soil creep, debris slides, slumps, etc.; see Figure 1-4); 6 = deep-seated mass movements (earthflows, etc.; see Figure 1-4); 7 = nutrient uptake. Not shown in figure: volatilization, wind erosion, nitrogen fixation, denitrification. Although it is not shown, understory vegetation also contributes to these processes.

This report discusses the implications of spatial scaling in forest hydrology and management. Spatial scale terms used in this report are defined in Box 1-1. Forest hydrology studies are conducted in plots, small experimental watersheds, and across landscapes and regions. Process studies and modeling are most commonly conducted at the small watershed spatial scale. At various scales, these process-based studies are used to examine the mechanisms of energy, water, sediment, and nutrient movement and transformations. Temporal scales of forest hydrology studies range from days to multiple decades, but many studies examine periods from a single storm event to a few years.

The first paired watershed experiment in North America to quantify the hydrologic effects of forest management was conducted by the USFS from 1909 to 1928 in southern central Colorado (Bates and Henry, 1928). By the 1960s, the USFS had established more than 100 experimental forests and experimental watersheds in the United States (USDA Forest Service GTR NE-321, 2004), and other public agencies, universities, and private companies established additional small watershed studies around the world (Ice and Stednick, 2004). Fifty years ago, more than 150 experimental watersheds were being studied in the United States, but only a handful of those are still active today (Ziemer, 2000). These small watershed studies are the foundation of our current understanding and predictive capabilities of the effects of forest harvest practices on runoff.

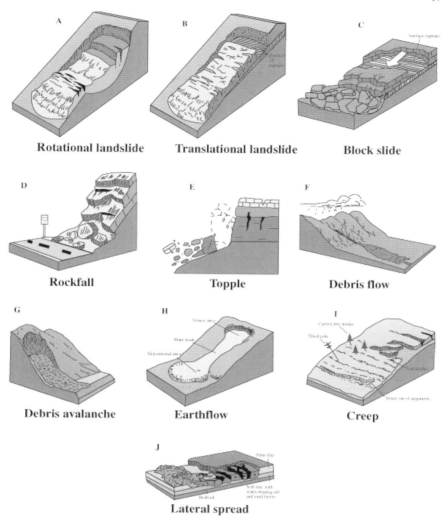

FIGURE 1-4 Mass movement processes in the forest. SOURCE: USGS (2004).

> **BOX 1-1**
> **Definition of Spatial Scales**
>
> **Plot scale:** areas of 10^0 to 10^2 m
> **Small experimental watershed:** drainage area up to 5 km^2
> **Large watershed:** drainage area up to hundreds of square kilometers that drains to a reservoir or lake that is part of the water supply infrastructure
> **Landscape:** collections of several large watersheds
> **Region:** multiple municipal areas, each of which has its own water supply

In the early twenty-first century, water and resource managers are asking questions that challenge forest hydrologists to go beyond general principles and study designs of the past to make predictions and respond to emerging issues. These include, for example, questions about cumulative watershed effects in large watersheds, legacy effects of roads on peak flows and sediment movement, or direct and indirect effects of climate change on forest hydrologic processes. The present body of knowledge provides a foundation for answering these questions, but there are significant information gaps and research needs, described later in this report (see Chapters 3 and 4).

These issues and questions are the centerpiece of the tensions in basins around the country. Scientists, managers, and the citizenry are looking for new approaches to more fully understand watersheds, make stronger connections between forests and water, and achieve multiple stakeholder goals.

THE NRC STUDY OF HYDROLOGIC EFFECTS OF FOREST MANAGEMENT

The Department of the Interior Assistant Secretary for Water and Science initiated discussions in 2005 with the Water Science and Technology Board (WSTB) of the National Academies' National Research Council (NRC) for an assessment of the science of forest hydrology and how it relates to hydrologic effects of forest management practices. The USFS joined these discussions at the end of that year. Together, the U.S. Bureau of Reclamation and the USFS requested that the WSTB convene a committee to produce a report on the comprehensive understanding of forest hydrology, connections between forest management and attendant quality and quantity of streamflow, and directions for future research and management needs. In early 2006, the WSTB formed the Committee on Hydrologic Impacts of Forest Management, a panel of 14 members with expertise in forest hydrology and ecology, fire ecology, watershed sciences, geomorphology, water quality, and forest management on public and private land ranging from small woodlots to extensive industrial holdings (see Appendix B). The overall charge to the NRC committee was to examine the effects of forest management on water resources (see Box 1-2). The committee held five meetings between March 2006 and April 2007 in open and closed sess-

---

BOX 1-2
Statement of Task

This study will examine the effects of forest management on water quantity, quality, and timing. The report will reflect on the state of knowledge, relevant policy implications, and research needs that would advance understanding of connections among hydrology, science, and land management and policy in forested landscapes.

1. What is the state of knowledge of forest hydrology?
2. What are information and research needs regarding forest hydrology in forested lands?
   - Topics could include: sediment-related watershed processes, surface and groundwater hydrology; biological and ecological aspects; and extrapolation of small-scale study results to large-scale management practices.
3. What are the new issues that need to be addressed to ensure clean and plentiful water?
   - Topics could include: extreme weather events, climate change, fire, and invasive species.
4. How well are forest hydrologic impacts understood over short- and long-temporal scales and small- and large-spatial scales?

---

ions around the United States to gather information and examples for this report and to hear perspectives from forest managers, water supply system managers, and water users on key issues related to forests and water.

Scope of the NRC Study

The committee produced this report to have maximum application and utility for a diverse audience of scientists, forest and water managers, and citizens in the community. To best reach this broad audience, the committee clarifies three points in its interpretation of the statement of task. First, this report expands the focus to be applicable to state and private forests, in addition to forested lands under federal management. Second, the report takes a national view of issues related to forests and water. The federal sponsors of this study have land holdings and jurisdiction primarily in the western United States, but issues and concerns about water and forests are evident in all 50 states. Finally, this report provides recommendations for scientists, managers, and citizens on approaches that can begin to ease tensions over water resources. Given the wide interest in the array of issues associated with forests and their hydrologic effects, this report builds on decades of forest and forest hydrology research to present key findings and recommendations that advance the understanding of connections among forests, water, and people and make that understanding accessible to scientists, managers, and citizens.

Structure of the Report

The following four chapters of the report describe the current understanding of forests and water, discuss information gaps and research needs in forest hydrology and management; and present recommendations to address issues and challenges in the science, research, and management of forests and water. The descriptions and discussions of forests (Chapter 2) include the primary management objectives, ownership patterns, and historic and emerging issues in forests. The state of the science of forest hydrology and the understanding of how forest management activities affect streamflow quantity and quality are assessed and presented in Chapter 3, including general principles and basic processes that have been gleaned from the forest hydrology literature. Research needed for managing forests and water in response to contemporary challenges is discussed in Chapter 4, with an emphasis on moving from principles to prediction at larger spatial scales and longer time scales. The report's final chapter (Chapter 5) draws upon the state of the science (Chapter 3) and research needs (Chapter 4) to make recommendations for scientists, managers, and communities to meet forest and water needs in this and future generations.

2
Forests and Water Management in the United States

Forests and water are inextricably connected. Forests process the water that sustains agriculture, human settlements, and ecosystem functions. Forests vary due to differences in geography; ecology; and social, economic, and land use histories. Throughout the United States, forests are managed for a range of objectives and goals, using a wide variety of forest management practices, which are regulated by diverse laws at the federal and state levels. Like forests, water resources are managed to achieve multiple objectives and are constrained by various laws. These laws and institutions fragment the management of forests and water, despite their close physical and biological connections. The variations in forest types, regions, objectives, and management, combined with the fragmented management of forests and water, creates a new body of emerging issues for forest and water managers.

This chapter describes current practices, past legacies, and future issues related to how forests are used and managed in the United States. It describes the regional differences in the relationships between forests and water; outlines forest management and water resource management objectives and practices; and examines how ownership patterns, laws, regulations, and institutions govern the use and management of forests and water. Finally, this chapter introduces the emerging issues for water and forests relevant to forest hydrology science and management in the twenty-first century.

FORESTS AND WATER IN THE UNITED STATES

The forests of the United States are diverse. They differ in regional characteristics and values, species composition, and forest types (Figure 2-2) and in ownership and management objectives. Forests account for 33 percent of all U.S. land area (Figure 2-3), covering about 750 million acres (300 million hectares) (Powell et al., 1993; Smith et al., 2004). Of this, 57 percent are privately owned, and the remainder is public forest. Ten percent of U.S. forests cannot be harvested for commercial timber because they are in areas designated as wilderness, parks, and other legally reserved classifications. Although the federal government owns or manages land in all 50 states, the vast majority of federal forest land is concentrated in 13 western states (Figure 2-3).

The geography, ecology, economics, and land use histories of forests differ markedly by region (Smith et al., 2004). More than half of forest area in the United States lies east of the Mississippi River (Figure 2-1a). In eastern forests, precipitation exceeds evaporation and transpiration on an annual basis, provid-

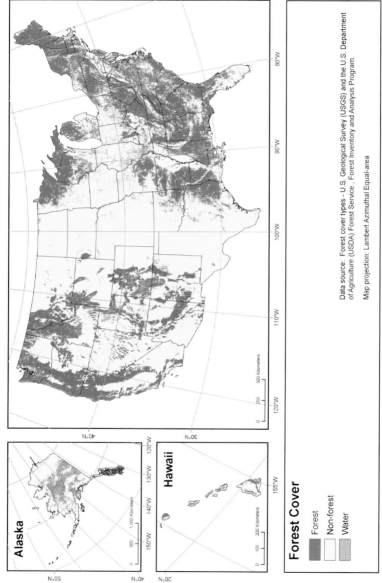

FIGURE 2-1 Forest cover in the United States. SOURCE: Map created by Catchment Research Facility, University of Western Ontario.

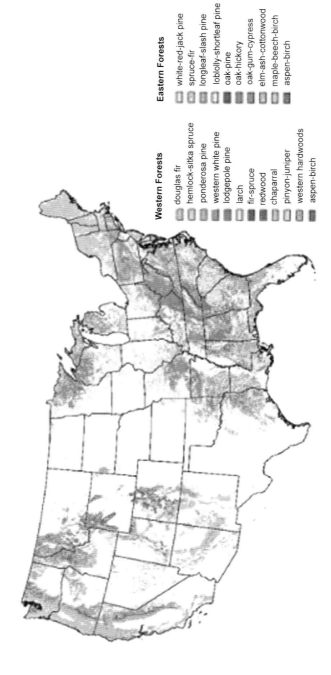

FIGURE 2-2 Map of forest vegetation types for the United States. SOURCE: USDA Forest Service.

26

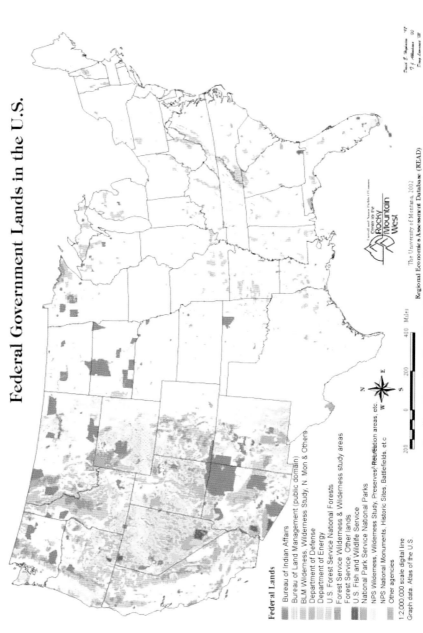

FIGURE 2-3 Federal lands in the United States. SOURCE: Reprinted, with permission, from Regional Economics Assessment Database (2002). Copyright 2002 by the Regional Economics Assessment Database.

ing abundant water supplies (Figure 3-2a). Northern and northeastern forests contain a mixture of broad-leaved (deciduous, hardwood) and conifer (evergreen, softwood) tree species, and slightly more than one-half of southern forests are conifer species (Figure 2-1b). The vast majority of eastern forests are second-growth that regenerated after land conversion to agriculture one to three centuries ago, followed by subsequent farm abandonment (Williams, 1989; Foster and Aber, 2004). More than 85 percent of the area of eastern forest is privately owned (Smith et al., 2004); compare Figures 2-1a and 2-1c.

The remaining half of forest area in the United States is located in the Rocky Mountain, southwestern, Pacific Coast, and Alaska regions (Powell et al., 1992; Smith et al., 2004) (Figure 2-1a). Many western forests have been shaped less by human disturbances than by natural disturbance, especially wildfire. Almost three quarters of western forests are on public lands (Smith et al., 2004; compare Figures 2-1a and 2-1c). Western forests are dominated by conifer species, but aspen, oak, and riparian forests are important broad-leaved components (Figure 2-1b).

Regional differences lead to contrasting relationships between forests and water in the West compared to the East. Precipitation in the West is strongly related to elevation, and at higher elevations most precipitation occurs as snow. In the West, most precipitation falls on forested mountains that are sparsely populated and on public lands, and these headwater areas provide the source water for public and private water supply systems that store, transfer, and deliver water to farms, people, and industry. In the East, headwater sources are often on private land, closer to end users, more densely populated, and containing a wider range of land uses than in the West; interbasin transfers are less common. As a result, issues involving forests and water in the West often focus on allocation of scarce water and involve federal agencies, whereas in the East, issues have focused on pollution and involve private as well as public land owners.

MANAGING FORESTS AND WATER

Forests and water are connected by physical and biological processes, so the management of forests affects the quantity, quality, and timing of water (Anderson et al., 1976; Waring and Schlesinger, 1985; Ice and Stednick, 2004). Federal laws and forest ownership influence the goals and objectives of forest and watershed management, which in turn determine management practices. Forest and water management is fragmented among many laws and institutions.

Forest Management Objectives and Practices

Forest management applies biological, physical, social, economic, and policy principles to meet specific goals and objectives. Forest management is a balancing act among the various uses and products. Forest management objectives can encompass producing timber for wood products; protecting or enhancing flows of high-quality water; providing herbage (forage) for livestock or other herbivores; enhancing food, cover, and water for wildlife habitats; or creating landscapes for outdoor recreational values. Additional forest management objectives include sustaining forest ecosystems; preventing or mitigating wildfires or insect outbreaks; preserving habitat for native species and combating the spread of invasive species; and conserving biological diversity.

Historically, many forest management practices have centered on timber management. Timber management encompasses silvicultural treatments to establish and sustain wood production; protection against or control of wildfire occurrences, insect infestations, and diseases; and of course, harvesting of merchantable trees in a forest. Silviculture, forest protection, and timber harvesting involve a number of actions that individually and cumulatively can modify water quantity, quality, and timing. Silvicultural practices include selection of species and genotypes, site preparation, planting, drainage, fertilization, watering, herbicide application, and thinning to maximize the growth of the most desirable species. Forest protection practices include fuel reduction treatments such as overstory thinning, understory removal, or prescribed fire; construction of fire breaks and fire lines; applications of soil, water, or fire-retardant chemicals; application of insecticides and fungicides; and introduction of biological control agents. Timber harvest practices include selection of the rotation age, which determines the ranges of forest ages; road and trail construction, including road drainage systems such as culverts; felling and skidding of logs to landings; and movement of logs, usually by trucks, to timber mills for processing.

A number of laws and regulations govern the lands managed by the U.S. Forest Service. These laws stipulate how the effects of forest management on watersheds must be addressed in management plans and actions. The 1960 Multiple Use and Sustained Yield Act (16 USC 525-531) recognized that national forests are important watersheds. Water diversions and associated ditches, pipelines, and canals are authorized on national forests through the issuance of special use permits or the granting of rights of way. Since the 1970s, the Forest Service also has appropriated water resources and asserted water rights to protect instream flows for fish habitat and outdoor recreation (Wilkinson and Anderson, 1985).

Public concern about adverse effects of clear-cutting for timber production led to passage of the National Forest Management Act in 1976 (16 USC 1604(g)(3)(E)). In management plans required by the Act, the Forest Service must ensure that timber is harvested only where soil, slope, or other watershed conditions will not be irreversibly damaged. The act also requires that timber harvest plans protect stream systems and stream banks, lakes and shorelines, wetland systems, and other bodies of water; prevent detrimental alterations in water temperatures; and limit sediment contributed to stream channels (see Box 2-1).

States, municipalities, and counties own and manage forestlands for various goals and objectives, and they often give broad discretion to a specified organi-

> **BOX 2-1**
> **Aquatic Conservation Strategy in the Northwest Forest Plan**
>
> Many national forests have adopted specific policies to protect water and watersheds beyond the basic requirement of the National Forest Management Act of 1976. The Aquatic Conservation Strategy addresses water and watershed protection as part of the Northwest Forest Plan (USDA and USDI, 1994), which governs timber harvest on federal lands in the Pacific Northwest. This strategy is a part of the land and resource management plans for each national forest and Bureau of Land Management district in the area. Unlike conservation and management plans of the past, the Aquatic Conservation Strategy addresses the entire riparian ecosystem over a large landscape. It seeks to prevent further degradation of aquatic ecosystems and to restore and maintain habitat and ecological processes responsible for creating habitat over broad landscapes, as opposed to looking at the effects of individual timber sales (USDA and USDI, 1994).
>
> Implementation of the Aquatic Conservation Strategy (ACS) illustrates the challenge of managing forestlands at a watershed scale. Developers of the Aquatic Conservation Strategy recognized that periodic disturbances would occur over the many years needed to restore ecological processes and that short-term disturbances were critical for long-term aquatic ecosystem productivity. However, they did not expect all watersheds to have favorable conditions for fish habitats at any particular time. Implementation of the ACS brought major changes to the way the affected land management agencies viewed and managed aquatic resources and watersheds; the ACS changed the focus from small spatial scales (i.e., project areas) to larger landscapes. The implications of these changes have not been recognized fully or appreciated by the land management and regulatory agencies or the general public, and
>
> it has been difficult to implement this underlying scientific premise into forest plans (Reeves et al., 2006).
>
> Environmental plaintiffs used the objectives of the Aquatic Conservation Strategy to challenge individual timber sales. The forest plans were amended in 2004 to clarify that larger watersheds and long-term time frames are the appropriate level to evaluate progress toward these objectives, not specific projects (USDA et al., 2004). This decision was itself successfully challenged in Pacific Coast Federation of Fishermen's Association v. National Marine Fisheries Service (U.S. Dist. LEXIS 23645 [2007]), which underscores the difficulty of successfully moving to landscape-scale analysis.

zation or agency to manage the lands (Rice and Souder, 1998). For example, laws direct state forests in California to be managed for maximum sustained production of high-quality forest products, while giving consideration to values for outdoor recreation, watershed, wildlife, range and forage, fisheries, and aesthetic enjoyment. Some states have independent certification to manage their forests sustainably, meeting watershed protection standards beyond those required by state law. Municipal or regional water authorities frequently own or manage forestlands to protect and control the watersheds from which they get their drinking water. The scale of management of these lands varies greatly from medium-sized towns that own and manage the watershed immediately surrounding their reservoirs to large urban areas that own reservoir lands and manage other lands in their larger watersheds through agreements designed to protect water quality (Boxes 2-2 and 2-3).

Private forestlands are managed to meet their owners' goals and objectives. Private forest industry companies generally manage their lands to produce timber, pulp, or other wood products. Nonindustrial private forest landowners, including small family forests and Indian tribes, often manage their forests for a wider range of purposes than producing wood products, such as maintaining wildlife habitat conditions or providing opportunities for outdoor recreational experiences.

A major shift in forestland ownership in the private sector and a concomitant change in management goals and objectives are occurring among large vertically integrated forest product companies. The passage of the Employee Retirement Income Security Act in 1974 encourages institutional investors to diversify their portfolios, which encouraged many industrial landowners to sell portions of their forestlands to timberland investment management organizations or real estate investment trusts. Instead of traditional forestry goals and objectives of supplying wood products, the primary management goal of these new kinds of investors is their own financial return. This shift in ownership accelerated in the 1990s with major restructuring in the forest products industry in response to increasing globalization as companies consolidated to become larger or even transnational. Forestry, therefore, has emerged as a new asset opportunity for investors rather than an asset owned by manufacturing companies and small woodlot owners (Sande, 2002).

Watershed Management Objectives

Watersheds are managed to provide sustained supplies of high-quality water for human uses (Heathcote, 1998). Cities and towns depend on watersheds, many of which are forested, for their water supplies. Many watersheds that provide drinking water supplies include forests that are actively managed (e.g., Dissmeyer, 2000; NRC, 2000). Many controversies arise when forest management in municipal watersheds is viewed as being in conflict with watershed management goals (Boxes 2-2 and 2-3).

BOX 2-2
Forest and Watershed Management Conflicts for Oregon's Largest Cities

Like many communities in the Pacific Northwest, the cities of Portland, Eugene, and Salem, Oregon, depend on water supplies from surface water that originates in watersheds on which extensive forest management activities occur. These watersheds are predominantly forested, with federal lands in the upper basin above flood protection reservoirs and a mixture of private and state lands downstream of the reservoirs but above the municipal water intakes. Forest management activities on these watersheds include road building, timber harvesting, post-harvesting chemical treatments, fire suppression, and firefighting with flame retardants. The following examples illustrate the complex challenges faced by water managers in these forested watersheds.

continues next page

BOX 2-2 Continued

In Portland, the challenge has been to reconcile timber management and water supply protection objectives. Since the early 1900s, the city of Portland has obtained a major source of unfiltered drinking water from the Bull Run watershed, which drains the Mount Hood National Forest. The USFS implemented austere restrictions on the uses of and access to the watershed in order to protect the city's water source. Public entry to the watershed was prohibited, roads were paved to prevent sediment production, and even horses that were used for logging in the watershed were equipped with diapers. At the same time that public entry was restricted, the U.S. Forest Service continued its patch clear-cutting and salvage logging operations in the watershed from the 1950s to the 1980s, which caused tremendous public outrage. This outrage was exacerbated after extreme windstorms blew down forests along clear-cut edges in 1973; the USFS resumed salvage practices on the windthrown trees, creating new clear-cut edges; and further windstorms in 1983 blew down additional trees along the fresh clear-cut edges (Sinton et al., 2000). Public controversy over apparent risks to Portland's water supply led to unilateral cessation of clear-cut logging in the watershed in the late 1980s.

In Salem, the management challenge was a conflict between management objectives. The North Santiam is the sole source of drinking water for the city of Salem. In a major flood in February 1996, high turbidity in the river from private forestlands downstream of the federally managed reservoir caused the city of Salem to shut down its water supply system. Turbid water also was caught and held in the federally managed flood control reservoir in the upper basin, which drains federal forestland. Over a week following the flood, water releases from the flood control reservoir maintained high turbidity levels and kept Salem's water supply shut down (Bates et al., 1996; GAO, 1996). In this case, federal management of a reservoir for one objective (flood control) conflicted with the achievement of another objective (water quality).

In the city of Eugene, the challenge was incompatible management objectives. The McKenzie River is the sole source of unfiltered drinking water for Eugene. In the early 2000s, fisheries biologists judged that late summer releases of cold water from reservoirs in the McKenzie River drainage above Eugene were having deleterious effects on native fish populations. The U.S. Forest Service undertook a project to retrofit one of the flood control reservoirs in the basin with a temperature control tower in order to provide water releases whose temperatures would be suitable for downstream fish populations. During construction of the tower, the reservoir was drained, and sediment was mobilized within the lowered pool. This sediment contributed turbidity to the McKenzie River for a period of several months. In this case, federal management of a reservoir to meet one objective (water quality-temperature) compromised another objective (water quality-sediment).

BOX 2-3
Forest and Watershed Management Conflict in Massachusetts

In contrast to western U.S. cities, most cities in the northeastern United States derive their municipal water supplies from a combination of private and state-owned land. The Boston metropolitan area derives about 90 percent of its safe yield (300 million gallons per day) as an unfiltered water supply from the Quabbin Reservoir (*http://www.mass.gov/dcr/waterSupply/watershed/water.htm*). The Massachusetts Division of Water Supply Protection owns and manages 65 percent of the 486 km^2 watershed; other public forests account for 7 percent and private forestland accounts for 24 percent of the total area.

Conflicts arise over forest management in the Quabbin between groups that favor active forest management for timber production, wildlife habitat, and other values, versus conservation groups that favor forest preservation and natural disturbances. On the one hand, although many activities are strictly regulated in the watershed, the Massachusetts

continues next page

> **BOX 2-3 Continued**
>
> Division of Water Supply Protection has practiced active forest management in 186 km^2 of the Quabbin Forest since the early 1940s, harvesting about 4 to 8 km^2 per year. Management practices include timber harvest aimed to protect forests from episodic disturbances (e.g., hurricanes, severe ice storms) and chronic disturbances (e.g., insect and disease outbreaks, browsing by white-tailed deer, atmospheric deposition). Forest management in the 1950s and 1960s focused on reforestation (with nonnative red pine) following the 1938 hurricane, which blew down trees in many parts of the watershed (Foster and Boose, 1992). In the mid-1960s some red pine stands were converted to grassland maintained by mowing and prescribed burning aimed to augment water yield. Since 1985, silviculture, forest protection, and timber harvest practices have aimed to diversify the vertical structure, age class distribution, and species composition of the forest.
>
> On the other hand, conservation groups and forest ecologists in Massachusetts, including scientists at the Harvard Forest, advocate forest preservation for wilderness values in the Quabbin watershed. The 235 km^2 Quabbin Forest is the largest undeveloped forest area in Massachusetts. Conservation groups in the state have called for a cessation of logging to create a large "wildland" forest area (Foster et al., 2004). The Division of Water Supply Protection acknowledges that erosion from roads constructed for access to timber harvest and skid roads is a potential problem associated with timber harvest (Quabbin, 2007). Nevertheless, concerns about forest management effects on water quantity or quality appear to be secondary to arguments about forest conservation versus timber production in the debate over forest management in the Quabbin reservoir.
>
> SOURCE: Quabbin (2007).

Fragmented Management for Forests and Water

Forest management occurs in the context of complex and highly fragmented laws, regulations, and social institutions that deal with land and water use. Water management also can be fragmented along its path from the precipitation falling on the land, to water flowing through stream systems, to human or other uses of the water and ultimately to the ocean. Since John Wesley Powell's *Report on the Lands of the Arid Region of the United States* (Powell, 1889), water researchers and policy makers have recommended an integrated approach to watershed management that organizes land and water management around hydrologic systems. Integrated management of forests and water at the watershed level involves many different public agencies, landowners, and a diversity of public and private stakeholder interests. Responsibilities and interests include management for drinking water supplies, flood control, reservoir operations for hydropower production, water for irrigation, fish and wildlife, and outdoor recreation. However, forest and water management remain fragmented (WWWPRAC, 1998; NRC, 1999).

Fragmentation of ownerships and interests combined with fragmented responsibility for managing and regulating forest management has made integrated management of forests and water at the watershed scale virtually impossible (Arnold, 2005). Institutional fragmentation exists at multiple levels, ranging from the goals of the laws, regulations, and institutions; to land ownership responsibilities and interests; to specified missions of the agencies responsible for management. Institutional regimes governing water use, water quality, and forestland use often evolved separately and frequently have different goals and objectives. Water use laws, regulations, and institutions usually focus on the use of water out that has been removed from water bodies, rather than the roles of water in streams, lakes, and other bodies of water. Water quality laws, regulations, and institutions are structured to help stakeholders control the impacts of land uses, including the effects of forest use and management practices on the quality of available water resources. Together, institutions and regulations specify how forests will be used and managed, setting some forests aside for timber production, some for watershed protection, and still others for preservation.

Fragmentation of administrative responsibility for the effects of forest management on the hydrologic processes in watersheds and landscapes occurs both vertically and horizontally. Responsibilities are split vertically among various levels of government—federal, state, regional and local, and they are split horizontally within each level among agencies focusing on specific resources. One agency manages forestlands, another manages water resources, and a third regulates the impacts of forest management on water resources. Rarely does only one agency or manager control forest management and use across entire watersheds or landscapes. A further discussion of the institutional governance of water is presented in Appendix A.

EMERGING ISSUES FOR FORESTS AND WATER

A number of issues pose challenges for science to explain and predict the effects of forest management on hydrology. These issues can be grouped into the spatial, temporal, and social contexts of forests and hydrology; they are described in those contexts below and expressed with key questions, which are addressed in subsequent chapters of this report.

Spatial Context

Timber Management Practices

Many forests continue to be managed for timber production. In those that are, large areas of some uneven-aged, native forest are converted to even-aged, managed forests. Selective cuttings continue in many uneven-aged forests. For decades, forest hydrology research has focused on how forests can be managed without adversely affecting stormflows, erosion, and adverse changes in water quality (Bosch and Hewlett, 1982; Brooks et al., 2003; Chang, 2003; Ice and Stednick, 2004). Nevertheless, issues remain about how much and which types of forest management can be practiced in a watershed while still maintaining water quantity and quality (see Boxes 2-2, 2-3, and 2-4).

Question: What are the magnitude and duration of hydrologic effects due to timber harvest?

Riparian Ecosystems

Removal of forests and other streamside vegetation within riparian corridors degrades stream ecosystems and reduces populations of aquatic organisms (Rinne and Minckley, 1991; Rinne, 1996; DeBano and Wooster, 2004). In the 1970s and 1980s, these findings and resulting public concern led to increased protection of streamside vegetation and riparian zone restoration as part of many forest management plans (Macdonald and Weinmann, 1997; Verry et al., 2000; Baker et al., 2004). Very wide riparian buffers, such as those required by the Northwest Forest Plan (USDA and USDI, 1994), occupy large portions of forest area. Riparian zones also contribute wood to streams that may be mobilized during floods, potentially exacerbating downstream flooding (Box 5-3).

Question: What are the hydrologic effects of removing or retaining riparian forests over the long term and in large watersheds?

Cumulative Watershed Effects

Changes in forest cover and extent within a watershed can result from forest fragmentation (the subdivision of large, continuous forest patches into smaller, discontinuous patches); conversion from forest to developed uses; timber harvesting; and forest loss due to fire, disease, grazing, and insects. Cumulatively, the hydrologic effects could be considerable when assessed at the large-watershed or landscape scale. "Cumulative watershed effects" (CWEs) are the response to multiple land use activities that are caused by, or result in, altered watershed function (Reid, 1993; MacDonald, 2000).

Question: What are the CWEs of forest cover loss in large watersheds?

Temporal Context

Past Timber Management

Clear-cutting was historically a widespread timber harvest practice on public forestlands. Today, little clear-cutting occurs on public forestlands, and harvest of old growth rarely occurs. However, two aspects of past timber management have created legacies in present-day forests that may affect water. One of these is the legacy of even-aged forest stands created mostly by patch clear-cutting done in the twentieth century. As they grow, these young stands use water, but there is uncertainty about how much water is used compared to the older, native forest stands they replaced. A second legacy is the edges created by past clear-cutting, which may be susceptible to windthrow disturbance. Major windthrow events can augment flammable material in forests (Moser et al., 2008) and contribute to insect outbreaks (Powers et al., 1999), with potential direct and indirect effects on water.

Question: How do past forest cutting patterns affect water quantity and quality?

Past Grazing and Predator Removal

Forestlands, especially western forests, were managed as public grazing lands in much of the twentieth century. Grazing of domestic cattle and sheep led to reductions in forest cover, soil compaction and erosion, increased overland flow, and sediment contributions to streams on many public forestlands. At the same time, eradication of native predators (wolves, cougars, etc.) led to increased populations of native grazers, such as elk and deer. Largely resulting from changes in USFS and Bureau of Land Management regulations, the grazing of domestic animals on national forests declined in the late twentieth century, while efforts to reintroduce predators have had some effect on native grazer populations and behavior, especially in national parks. Forest vegetation responses to reduced grazing pressure and the resulting effects on water use are largely unknown.

Question: How have changes in grazing of both domestic and native grazers affected forests, and what are the indirect effects of those changes on water quantity and quality?

Inherited Road Networks

One legacy of past timber management is a 386,000-mile (620,000 km) road network on USFS land. These road networks were designed and constructed to meet forest management goals such as timber extraction, fire control, or recreational activities. However, many forest roads have been maintained infrequently and do not meet current road standards (Bell, 2000). Some portions of the road network have been decommissioned, but most of the original road network remains. Forest roads are major sources of landslides (Swanson and Dyrness, 1975; Megahan et al., 1978; Sidle et al., 1985; Sidle, 2000) and sediment loads in the streams originating in national forests (e.g. Reid, 1993; Wemple et al., 2001). The road network has been implicated in flooding (Box 2-4)

BOX 2-4
West Virginia Flooding

In July 2001, several large, long-lasting thunderstorms passed over southern West Virginia. More than 6.5 inches (165 mm) of rain fell in a 100- or 500-year event. The resulting floods caused extensive property damage, personal injury, and death. A total of 489 private residential property owners filed lawsuits against 78 different defendants, including logging companies, alleging that timber harvesting, mining, and other resource extraction caused or contributed to the flood damage by eroding the soil and making storm runoff more intense.

The damage claims are based on several theories of liability including strict liability; unreasonable use of land; negligence; interference with riparian rights; and nuisance. The state Supreme Court has rejected the strict liability theory, but the suits are proceeding on the other grounds. The plaintiffs claimed that timber companies were not following best management practices and that they carried out their harvesting, road building, and other activities in ways that unreasonably harmed downstream landowners by increasing the intensity of the runoff, eroding soil, and carrying logs downstream into buildings and other structures (Mortimer and Visser, 2004).

A Flood Protection Task Force was formed by the governor and it prepared a new comprehensive Statewide Flood Plan (*http://www.wvca.us/flood/*). The task force concluded that while forest harvesting operations may affect flood flows due to soil compaction, the major flooding risk associated with logging relates to road and culvert design and maintenance. The task force recommended increased inspections of forest operations, prompt reforestation, and improved management of logging slash.
---

and represents a potentially very large future source of sediment, which could adversely affect water quality in forested watersheds.

Question: How do the legacies of road networks on forestland affect peak flows and sediment movement?

Wildfire and Fire Suppression

Fires are a natural disturbance in many forest ecosystems. Natural fires (or wildfires) range in recurrence-frequencies of less than 10 years in ponderosa pine forests in the Southwest to more than 1,000 years in balsam fir forests in the eastern United States (Swetnam and Baisan, 1996; Swetnam, 2005). Depending on their severity, wildfires may affect energy and nutrient flows, the soil environment, above- and below-ground plant growth, wildlife populations and their habitats, and hydrologic processes. For most of the twentieth century, wildfires were effectively suppressed to protect timber. As a result of fire suppression, flammable fuels have accumulated in many western forest ecosystems. These fuel accumulations are believed to contribute to increased wildfire size and severity (see Box 2-5). Forest managers have begun to reintroduce fire to some national forests using prescribed fire and "let-it-burn" policies (Schullery, 1986; Arno and Brown, 1991; Czech and Ffolliott, 1996). However, firefighting

> **BOX 2-5**
> **Historical 2002 Wildfire Season and Lessons Learned**
>
> More than 88,000 wildfires burned throughout forest and rangeland ecosystems of the United States in the 2002 wildfire season that burned a record near-7 million acres of forestland—almost twice the 10-year average. Aided by a widespread drought in the western regions that was comparable in severity to the Dust Bowl of the 1930s and the excessive buildup of flammable fuels, these wildfires initiated a heightened public awareness of the dire consequences of large and complex wildfires (White, 2004). Three historically large wildfires caused by human ignitions—the Rodeo-Chediski in Arizona, the Hayman in Colorado, and the Biscuit in southern Oregon and northern California—burned more than 1 million acres combined and placed the impacted ecosystems, communities, and people at risk. Given the inevitability of wildfires in the future; the escalating impacts of wildfires on natural, human, and economic resources; and the need for improved preparation to combat future wildfires, the experience from the 2002 fire season provided important lessons for foresters and fire managers:
>
> - The need to improve knowledge of and ability to predict the risk of wildfires occurring in a particular locale;
> - The need to develop plans for the recovery of ecosystems to fire; and
> - The need to estimate and plan for the direct costs involved in mitigating wildfires and rehabilitating burned landscapes.

remains a major forest management practice of the Forest Service on most national forests in the western United States (see Box 2-6). The Forest Service's use of fire retardant chemicals is a source of controversy and concern (see Box 2-6).

Questions: What are the hydrologic effects of forest fires and firefighting (such as fire breaks, soil disturbance, and application of fire retardants)? What are the hydrologic effects of high versus low-severity fires, including considerations of long-term effects and larger spatial scales?

Impacts of Insect Outbreaks

Under normal conditions, bark beetles, leaf defoliators, and other insects are present at low (endemic) levels in forest ecosystems. However, when conditions are favorable, endemic populations erupt into epidemics, altering forest species' composition and structure and killing all of the trees in severely infested forest stands. In the early 2000s, much of the western United States has been experiencing bark beetle outbreaks at unprecedented levels, apparently as the result of warming climate (Bytnerowicz et al., 1998; Logan et al., 2003). Vast areas of western forest have been killed by these large-scale outbreaks, especially in the Rocky Mountains.

> **BOX 2-6**
> **Judge Threatens to Block Forest Service Fire-Retardant Drops and Put Government Official in Jail**
>
> A federal judge in Montana is threatening to block the Forest Service's use of fire retardant drops and throw the Agriculture Undersecretary in jail. In January 2008, the U.S. District Judge ordered the Forest Service to court to explain why the agency has failed to conduct proper studies of fire retardant drops in Missoula. If the agency's arguments are unpersuasive, the judge said he would consider enjoining the use of all aerial fire retardants nationwide, except for water, until the Forest Service complies with his orders and federal environmental laws.
>
> At issue is firefighters' use of fire retardants containing ammonia compounds. Federal and state agencies drop an average of 15 million gallons of retardant annually, up to 40 million gallons in some years. Fire retardant is approximately 85 percent water, but it contains ammonia compounds, thickeners such as guar gum and attapulgite clay, dyes, and corrosion inhibiters. Retardant rapidly reduces wildfire intensity and rate of spread by robbing the fire of oxygen and slowing the rate of fuel combustion with inorganic salts.
>
> The Forest Service Employees for Environmental Ethics (FSEEE) initiated the lawsuit in 2002, alleging that fire retardants are used without analysis of their environmental impacts and blaming fire retardants for fish kills, including the death of 20,000 fish in central Oregon in 2002. The lawsuit alleges that the USFS is violating the National Environmental Policy Act (NEPA) and failing to comply with Section 7 of the Endangered Species Act, which requires the agency to consult with the National Marine Fisheries Service and the Fish and Wildlife Service. A biological opinion from the National Marine Fisheries Service found potential harm to 24 threatened and endangered fish species in the Northwest, including nine species of chinook salmon, two species of chum salmon, two species of coho salmon, Snake River sockeye salmon, ten species of steelhead, the shortnose sturgeon, and the green sturgeon. In October 2007, the Forest Service made an initial finding of no significant impact.
>
> FSEEE sees the lawsuit as a way of making the Forest Service rethink traditional firefighting strategies that now consume half of the its budget, even as many national forests can no longer afford to maintain their campgrounds and trails. "Fire retardant is the wedge we're using to force a hard look at the way we fight fires—how we fight them, where we fight them, when we fight them, and why we fight them," said Andy Stahl, executive director of the group. The Forest Service's fire suppression budget has roughly doubled since 2000 as fires have grown more intense. The environmental group wants the Forest Service to treat fire retardants in the same manner as pesticides, which the agency uses infrequently and in limited areas, generally after much study.
>
> SOURCES: From the Oregonian, February 26, 2008, *http://www.fseee.org/fsnews/ee080111.pdf*. Available online at *http://blog.oregonlive.com/pdxgreen/2008/02/top_bush_official_faces_jail_f/print.html*.

Questions: How do insect outbreaks affect water quantity and quality? How can future hydrologic effects of insect outbreaks be understood or predicted as indirect effects of climate change?

Spread of Invasive Species

The effects of invasive plant species on forest ecosystems are a major concern of foresters and ecologists (Young and Clements, 2005; Webster et al.,

2006). At least one invasive species was found in 62 percent of the 200 forested plots studied in Oregon and Washington (Gray, 2007). Invasive species in forests include grass, herb, shrub, tree, insect, and bird species that are exotic (nonnative), as well as native species that spread beyond their historic range with undesirable impacts (see Box 2-7). Invasive plant species can displace native plants; modify habitat for native insects, birds, and animals; and alter ecosystem processes including nutrient cycling, fire, and water use (Flather et al., 1994; Wilcove et al., 1998; Callaway and Aschehoug, 2000). Eucalyptus, Russian olive, and tamarisk are common invasive tree species in western forests, and ailanthus and kudzu (a vine) affect eastern forests.

Various forest management practices exist for control of invasive species (Wagner et al., 2000; Falk and Swetnam, 2003; Hull Sieg et al., 2003). The Forest Service, along with nongovernmental organizations such as The Nature Conservancy, has adopted practices to inventory and limit the introduction and spread of invasive species on many national forests and other forested lands.

Question: What are the hydrologic effects of nonnative species' presence and nonnative species' removal treatments in forests?

Changing Climate

Changing climate is directly influencing forest ecosystems, with consequent effects on water quantity, quality, and timing of peak and low flows. Effects of climate change on forests involve interactions among increasing CO_2, warming, changes in precipitation regimes (Melillo et al., 1993; Houghton, 1995), and the abilities of plant and animal species to migrate and keep pace with climate change or adapt to the changing conditions (Thomas et al., 2004). Ecological models incorporating alternative climate scenarios indicate that the location and extent of potential habitats for many tree species and forest ecosystems are likely to shift (U.S. Global Climate Research Program, 2000). Habitats for tree species favoring cool environments might move north or higher in elevation. Habitats of alpine and subalpine spruce-fir forests on isolated mountain tops in

---

BOX 2-7
Invasive Plant Species: Some Characteristics and Features

Timber harvest, road construction, fire, and grazing produce soil disturbance than can promote the establishment of invasive plant species in forests (Flather et al., 1994; Hull Sieg et al., 2003). Once introduced, invasive plants can be spread throughout a forest by traffic along road and trail networks (Parendes and Jones, 2000), potentially reaching sites that have experienced little or no human disturbance. Once they are present in a forest, invasive species also can be spread by natural disturbance, such as floods and wildfire (Watterson and Jones, 2006). Invasive species typically have life history traits that favor rapid establishment and spread, including high rates of seed production, edible seeds, vegetative reproduction, and persistence in the soil seed bank.

the southwestern region of the country could be eliminated if future elevational shifts are large enough. Aspen and eastern birch forests might contract in area in the United States and shift into Canada. Other ecosystems might expand in area, such as oak-hickory and oak-pine habitats in the eastern states and ponderosa pine forests and pinyon-juniper woodlands in the western states.

Climate change will likely affect the yield, timing, and quality of water flowing from forest landscapes. Attempts to model forest responses to climate change and consequent water yield changes suggest a trend of declining water yields (Running and Nemani, 1991; Aber et al., 1995). A warming climate will reduce snowpack amounts and duration, and snowmelt will occur earlier. In some regions, spring peak runoff has already been documented as coming up to three weeks earlier than historical averages (Hodgkins et al., 2003; Dettinger et al., 2004; Payne et al., 2004). Even with conservative estimates of climate change, water resources to meet current demands are not guaranteed under future climate scenarios (Barnett et al., 2004).

Climate change also may alter the frequency and magnitude of forest fires, increasing the size and severity of wildfires (see Box 2-5). The effects of climate change on fire susceptibility involve complex interactions among factors such as warming temperatures, soil moisture, forest growth, fuel loads, and fire ignitions (Prentice et al., 1993; Stocks et al., 1998). Western forests are already experiencing larger and more severe fires and longer fire seasons (Kasischke et al., 2006; Westerling et al., 2006).

Questions: What are the hydrologic responses to climate change?

Social Context

Forestland Ownership Changes

Most regions of the United States have experienced a major transformation in private forest ownership during the past 20 years. In contrast to the twentieth century, there are very few remaining large, publicly traded, vertically integrated wood products manufacturing businesses that own significant amounts of forestland. Forested land and mills are increasingly owned by two new forms of large, privately held companies that are referred to as timber investment management organizations and real estate investment trusts. These companies now own what used to be industrial timberlands. They are investing in forestland ownership for the long term, but they have different time horizons and goals for management compared to the former owners of these lands.

At the same time, many family-held forestlands have undergone parcelization, which occurs at the time of intergenerational transfer when a forest changes ownership or is broken up and sold when the new owners cannot agree on goals and purposes. About one-half of the private forest land in the United States has changed ownerships in the past decade (Alig and Plantinga, 2004).

Question: How do changes in ownership affect forest management, and how do these changes affect water resources?

Continuing Urbanization

Urbanization has been a major cause of forest loss since 1950 and is anticipated to account for additional forest loss in the twenty-first century across the United States (Alig et al., 2003). Although forest area in the United States increased from the late 1800s through the middle 1900s, net forest area in the United States declined from the early 1950s to 1997 (Powell et al., 1993; Scott et al., 2004) (Figure 2-4). Forest area is expected to decrease by an additional 3 percent by 2050 relative to 1997 because of the conversion of forests to urban and developed uses. The U.S. Department of Agriculture's 1997 National Resource Inventory showed that 1 million acres of forest, agricultural cropland, and open space were converted to urban and other developed uses from 1992 to 1997, and the national rate of urbanization increased notably compared to the period from 1982 to 1992 (Figure 2-5). In the 1990s, forestland was the largest source of land conversion to developed uses. Aligned with a projected population increase of more than 120 million people through 2050, urban and other developments are expected to continue to grow substantially, with the fastest rates of growth in the western and southern regions (Alig et al., 2003). From 1990 to 2000, 18 states in the West registered growth above the U.S. average.

Continuing urbanization and increasing construction of second homes in forest settings has resulted in the expansion of "urban-forest interfaces" or "wildland-urban interfaces (Radeloff et al., 2005) throughout the country. Wildfires igniting in forests may spread into these communities and potentially be intensified if fuel buildup around homes is not managed (Cortner et al., 1990; Beebe and Omi, 1993; Vince et al., 2005).

Question: What are the effects of the expansion of human settlements into forested areas, and the consequent changes in forest management, such as thinning for fuel reduction, on water quantity and quality?

SUMMARY

This chapter describes current and emerging issues of managing forests and water in the United States. It describes forest management decisions that have been made in U.S. forests and enumerates, in the form of a set of questions, emerging issues that face forest and water managers in the twenty-first century.

The next two chapters address these questions. Chapter 3 evaluates how forest management and forest disturbance influence the flowpaths of water from precipitation to the point of human or ecosystem use. Chapter 4 identifies research needed to address these questions.

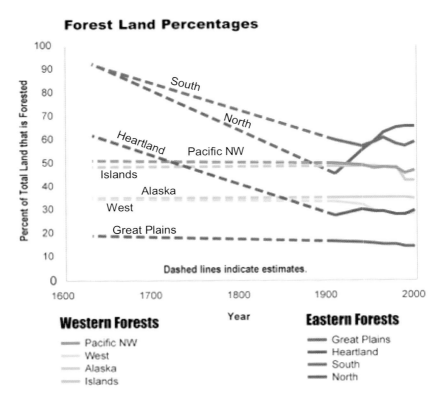

FIGURE 2-4 Change over time in forest cover by region of the United States. Forest cover declined greatly in the Northeast, South, and Midwest in past centuries, but in the twentieth century, forest cover declined in parts of the Pacific Northwest, and increased in the South, Northeast, and Midwest. In the twenty-first century, forest cover is projected to decrease in the South and Northeast to exurban development. SOURCE: Available online at http://www.usgcrp.gov/usgcrp/Library/nationalassessment/LargerImages/SectorGraphics/Forests/Percentages.jpg. Accessed June 24, 2008.

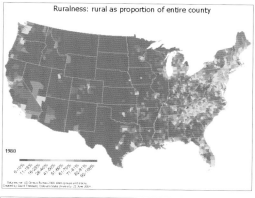
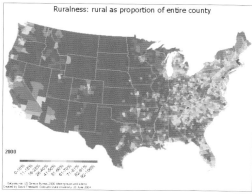
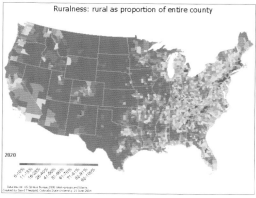

FIGURE 2-5 Proportion of rural area by county in 1980, 2000, and 2040. SOURCE: Reprinted, with permission, from Ecology and Society (2005). Copyright 2005 by David M. Theobald.

3
Forest Disturbance and Management Effects on Hydrology

Forest hydrology is the study of water in forests: the movement, distribution, and quality of water as regulated by forests. Forest hydrology addresses the hydrologic processes within forested areas and the output of water resources from forested areas. Disturbances—both planned and unplanned—and management activities in forests (see Chapter 2) can significantly alter hydrologic processes. These changes can affect nearly all components of forest ecosystems, including surface erosion, slope stability, nutrient cycling, channel morphology and aquatic organisms, and the quantity and quality of water.

This chapter defines forest hydrology, describes the factors that produce change in forests, and lists the general principles of forest hydrology (Figure 3-1). The chapter then provides an overview of forest hydrology findings to date and evaluates how this science supports the management of forests for water.

FOREST HYDROLOGY SCIENCE

Forest hydrology draws on forestry, including silviculture and forest watershed management, as well as civil, environmental, and hydraulic engineering; ecohydrology; geomorphology; soil science; and water resources engineering. It combines field measurements, experiments, and modeling to characterize and predict hydrologic processes and water resources. Principal instruments include precipitation and streamflow gages; devices for collecting and measuring water in tree canopies; thermocouples and other devices to measure sap flow in trees; wells, piezometers, lysimeters, and other devices to measure soil water tension, soil water content, and water table depth; and many types of devices to characterize chemical composition of water in trees, soils, and streams. The principal questions of forest hydrology are:

- What are the flowpaths and storage reservoirs of water in forests?
- How do modifications of the forest—including both trees and forest soils—influence water flowpaths and storage? and
- How do changes in forests affect water resources from forests?

Forest hydrology science relies on watershed studies, plot studies, process studies, and modeling. "Paired watershed" studies are an important approach to forest hydrology. In a paired watershed study, stream gages are installed at the mouths of two or more watersheds, and the watersheds are manipulated to determine the effects of experimental forest treatments on streamflow or water quality. Paired watersheds are similar in size, land use or land cover, and other

```
┌─────────────────────────────────────────────────┐
│            Modifiers of forest hydrology         │
│   Forest disturbance:      Forest management:    │
│       Wildfire          Forest harvest & silviculture │
│    Insects & disease         Road networks       │
│    Species changes              Grazing          │
└─────────────────────────────────────────────────┘
                        ↓
┌─────────────────────────────────────────────────┐
│      Hydrologic response: General principles     │
│  1. Changes in    2. Changes in    3. Changes in │
│     forest          flowpaths in     water, soil │
│     structure      soil and subsoil   chemistry  │
│  ┌───────────────────────────────────────────┐  │
│  │         Specific hydrologic responses      │  │
│  │      Hydrologic responses within forests:  │  │
│  │         Interception & transpiration       │  │
│  │          Infiltration & overland flow      │  │
│  │        Water flowpaths in soil and subsoil │  │
│  │                      ↓                     │  │
│  │         Changes in watershed outputs:      │  │
│  │                 Water yield                │  │
│  │                   Floods                   │  │
│  │                  Lowflows                  │  │
│  │                  Sediment                  │  │
│  │                  Chemistry                 │  │
│  │                 Temperature                │  │
│  └───────────────────────────────────────────┘  │
└─────────────────────────────────────────────────┘
                        ↓
         ┌───────────────────────────────┐
         │   Managing forests for water  │
         └───────────────────────────────┘
```

FIGURE 3-1 Forest hydrology examines the flowpaths and storage of water in forests, and how forest disturbance and management modify hydrologic responses. Hydrologic responses to changes in forests fall into three categories of general principles, as well as specific hydrologic responses, discussed in the text. The final section of this chapter evaluates the state of knowledge of forest hydrology and its implications for managing forests for water, including feedbacks to processes that modify forests.

attributes. After a reference period (usually five or more years), the "treated" watershed is subjected to manipulations (i.e., forests cutting, road building, fire, herbicides, etc.), but the "control" or reference watershed is not. The change in the relationship of streamflow and water quality between the treated and control watersheds before and after treatment is defined as the effect of the forest treatment. Most paired watershed studies in forest hydrology were begun in the 1940s through the 1960s, and many of these studies had been abandoned by the 1980s.

Field measurements from plots (segments of hillslopes), plot-scale experiments, process studies, and hydrologic modeling also are important components of forest hydrology. Field measurements from control and treated (e.g., burned, watered, cleared) forested plots have provided important insights into key hydrologic processes such as interception, transpiration, infiltration, and overland flow. Hundreds of hydrologic models have been developed to represent and predict water quantity and quality, and many of these models have been applied to forested areas; models are especially important for areas lacking field measurements and experiments.

In the twentieth century, forest hydrology science was conducted primarily in relatively small areas: segments of hillslopes ("plots") or small watersheds that ranged in size from a few square meters to 1-2 km^2. Time scales for plot-scale research commonly spanned a few growing seasons. Watershed studies have mostly been limited to one or two decades, but some, especially those in U.S. Forest Service (USFS) experimental forests and Long-term Ecological Research (LTER) sites sponsored by the National Science Foundation (NSF), have spanned periods up to 60 years. Most watershed studies occurred on publicly owned forestland, but some were conducted on privately owned land.

MODIFIERS OF FOREST HYDROLOGY

Forests are dynamic systems. Forests can be modified by: (1) fire; (2) species changes; (3) insects and disease; (4) forest harvest and silvicultural activities; (5) roads and trails, including skid trails; and (6) grazing (see Figure 3-1).

Hydrologic Effects of Fire

Record-breaking fires in recent years and projected increases in high-severity wildfires in the western United States (Fried et al., 2004; Westerling et al., 2006) have contributed to increased interest in how fire in forested systems affects water (Neary et al., 2005a). The type and magnitude of post-fire effects on runoff and water quality vary greatly with fire severity (Box 3-1), vegetation type, topography, soil type, subsequent amount and type of precipitation, and other local conditions. Despite these variances, fires can increase runoff and erosion rates by one or more orders of magnitude (Tiedemann et al., 1979; Helvey, 1980).

> **BOX 3-1**
> **Fire Severity and Hydrologic Effects**
>
> Most researchers and resource managers use fire severity as the primary means to characterize and predict the hydrologic effects of fires. Fire severity differs from fire intensity. Forest fire *severity* refers to the amount of tree mortality, which is related to fire effects on the ground surface. Fire *intensity* is the heat released per unit time per unit length of flame front.
>
> Fire severity is usually divided into three classes—low, moderate, and high. In low-severity fires the surface of the litter layer is blackened or partially consumed and some understory vegetation may be scorched or burned, but some charred or unburned litter is still present on the soil surface (Wells et al., 1979). In moderate-severity fires the litter layer is completely consumed, but the underlying mineral soil is not physically or chemically altered. High-severity fires not only burn all of the surface litter, but the greater soil heating also consumes some of the surface organic matter. This results in a disaggregation of particles in the uppermost layer of mineral soil and, depending on the soil type and amount of heating, a discoloration of the soil surface. In high-severity fires, more than 75 percent of the forest canopy is killed, and the forest floor may contain cavities where stumps and roots have been completely consumed as well as lines of white ash where coarse woody debris once covered the surface.
>
> High-severity fires typically have a much greater hydrologic effect than low- or moderate-severity fires. The effect of wildfires on life, property, and aquatic resources means that wildfire risk is a major factor driving forest management, particularly on public lands. Approximately 45 percent of the U.S. Department of Agriculture Forest Service budget is now being devoted to fire suppression and fuels management, which is nearly double the proportion allocated in the 1980s and 1990s.

Hydrologic Effects of Changing Species Composition in Forests

The composition of forest species changes as a result of natural disturbance and forest management. During forest succession after disturbance, early successional species are replaced by late successional species, leading to changes in water yield (Swank and Crossley, 1988; Jones and Post, 2004). Forest management may modify forest species, such as replacing deciduous forest with conifer species; these changes modify water yield and timing (Swank and Crossley, 1988). In addition, wildlife management, particularly the reduction of predators, may increase native ungulates and indirectly modify forest species composition by altering the intensity of browsing (Gill and Bearsall, 2001). The hydrologic responses to these indirect effects of wildlife management are not known.

Hydrologic Effects of Insects and Diseases

A wide array of insects and diseases can be found in forest ecosystems. Insect and disease dynamics are closely coupled to climate and forest disturbances such as wind, ice, and fire. The two groups of insects of primary concern to forest managers are bark beetles and defoliators (Schmid and Mata, 1996). Bark beetles bore into the living tissue and weaken or kill trees by introducing fungi and disrupting the transport of water and nutrients between the tree crown and the roots. Recent bark beetle infestations have killed extensive areas of mature trees in short periods of time. Defoliators generally lay their eggs in the buds of trees and feed on the emerging new leaves or needles. In contrast to bark beetles, defoliators take several years of repeated, heavy attacks to kill a tree.

Large, relatively infrequent outbreaks of insects or disease in forests are often a matter of public concern; they are also ubiquitous disturbance mechanisms in forest ecosystems. Low levels are termed *endemic*, whereas a rapid increase in insect populations or the incidence of disease is termed an *outbreak*. Historical photographs and tree ring records indicate that outbreaks of native species of insects and native diseases are intrinsic processes that are a natural part of forest ecosystems. However, introduced insects and pathogens are primary drivers of forest change and have transformed forests in the United States.

Large outbreaks of native insects and dramatic forest decline due to the introduction of exotic pests and diseases fueled early interest in the effects of insects and pathogens on forest hydrology (Bue et al., 1955; Bethlahmy, 1974). Recent increases in forest area affected by insect outbreaks and possible links to fire suppression (Fleming et al., 2002; Bebi et al., 2003) have reignited scientific interest in the effects of insect and pest outbreaks on water quantity and quality. However, very few studies have been conducted on the hydrologic effects of insects and disease. The hydrologic effects of insects and disease can be extrapolated from general principles derived from studies of timber harvest and fire (MacDonald and Stednick, 2003; Uunila et al., 2006), but much remains to be understood about hydrologic effects of insects and disease.

Hydrologic Effects of Timber Management and Silviculture

Since the 1950s and 1960s, when timber harvesting expanded on federal forest lands and industrial forestry developed on private lands, an extensive literature has examined the effects of forest management. Forest hydrology studies have addressed the effects of silvicultural practices (such as site preparation, herbicide treatment, and thinning); forest protection (such as post-harvest slash burning); and timber harvest (especially removal of trees, and construction of roads and trails) on water quantity and quality. Along with fire, insects, and disease, these are the primary processes that modify forests. Most of these studies have occurred in small plots or in small, experimental watersheds. Forest management effects on water quantity, quality, and timing vary with the area treated, the type of treatment, forest type, soils, climate, and time after treatment (Hibbert, 1967; Anderson et al., 1976; Bosch and Hewlett, 1982; Swank and Crossley, 1988; Hornbeck et al., 1993; MacDonald and Stednick, 2003; Jones and Post, 2004; Brown et al., 2005; Moore and Wondzell, 2005).

Hydrologic Effects of Roads and Trails

Forest management for timber and firefighting in the latter half of the twentieth century relied heavily on trucks and other heavy machinery for skidding logs to landings, constructing fire breaks, and hauling logs to mills. Starting in the 1940s, extensive road networks were constructed on public and private forestlands in the western United States, and heavy machinery was used on forest soils throughout the country.

Obvious soil disturbances associated with mechanized harvesting equipment and conspicuous landsliding associated with forest roads led to early interest in the effects of roads and trails on forest hydrology (Megahan, 1972; Anderson, 1974; Harr et al., 1975; Swanson and Dyrness, 1975; Ziemer, 1981). Continued hydrologic and sedimentation effects of lengthy road networks combined with efforts to decommission roads have fueled continued study of the effects of roads and trails on water quantity, timing, and quality (Reid and Dunne, 1984; King and Tennyson, 1984; Wemple et al., 1996; Bowling and Lettenmaier, 2001; Lamarche and Lettenmaier, 2001; Wemple et al., 2001; Wemple and Jones, 2003; Coe, 2006). Roads affect water timing and water quality, but the magnitude of the effect depends on road design, slope steepness, soils, and the configuration of the road system relative to the stream and river drainage networks.

Hydrologic Effects of Grazing in Forests

In 1970, about half (85 million acres) of the western forests and about four-tenths (161 million acres) of the eastern forests were grazed, and about one-half of the areas grazed in eastern forests was "exploitatitve," or beyond acceptable management (Anderson et al., 1976). Forest Service researchers (Platts, 1981) estimated that more than 800 million acres in the United States were grazed by livestock in 1970, furnishing 213 million animal unit months of forage. Overgrazing (animal densities in excess of the carrying capacity of the range) was common on forestlands and became a major research and management concern in the 1920s (Platts, 1981). Overgrazing in forests was associated with decreased infiltration capacity, increased overland flow and surface erosion, increased peak flows, and increased sedimentation in streams (Johnson, 1952; Dissmeyer, 1976; Anderson et al., 1976). In larger watersheds, overgrazing by domestic livestock is associated with ecological damage to thousands of linear miles of riparian forest cover and associated ecosystems, spurring policy statements by the American Fisheries Society (Armour et al., 1994).

HYDROLOGIC RESPONSES: GENERAL PRINCIPLES

Twelve general principles of forest hydrology (Table 3-1) describe the

TABLE 3-1 General Principles of Forest Hydrology Describing the Direct Effects on Hydrologic Processes of Changes in Forest Structure, Changes in Water Flowpaths, and Application of Chemicals

	Principles of Hydrologic Response to Changes in Forest Structure
1	Partial or complete removal of the forest canopy decreases interception and increases net precipitation arriving at the soil surface
2	Partial or complete removal of the forest canopy reduces transpiration
3	Reductions in interception and transpiration increase soil moisture, water availability to plants, and water yield
4	Increased soil moisture and loss of root strength reduce slope stability
5	Increases in water yield after forest harvesting are transitory and decrease over time as forests regrow
6	When young forests with higher annual transpiration losses replace older forests with lower transpiration losses, this change results in reduced water yield as the new forest grows to maturity
	Changes in Water Flowpaths in Soils and Subsoils
7	Impervious surfaces (roads and trails) and altered hillslope contours (cutslopes and fillslopes) modify water flowpaths, increase overland flow, and deliver overland flow directly to stream channels
8	Impervious surfaces increase surface erosion.
9	Altered hillslope contours and modified water flowpaths along roads increase mass wasting
	Hydrologic Response to Application of Chemicals
10	Forest chemicals can adversely affect aquatic ecosystems especially if they are applied directly to water bodies or wet soils
11	Forest chemicals (fertilizers, herbicides, insecticides, fire retardants) affect water quality based on the type of chemical, its toxicity, rates of movement, and persistence in soil and water
12	Chronic applications of chemicals through atmospheric deposition of nitrogen and sulfur acidify forest soils, deplete soil nutrients, adversely affect forest health, and degrade water quality, with potentially toxic effects on aquatic organisms

NOTE: These general principles are not predictions, so qualifying adjectives such as "may," "usually," etc., are omitted.

direct effects or first-order responses to changes in forest structure, changes in water flowpaths in soil and subsoil, and application of chemicals. These principles tie together the storage and movement of water in forests, how disturbance and management modify water storage and movement within forests, and how these internal changes are translated into changes in watershed outputs (Figure 3-1). These principles embody the state of knowledge of forest hydrology based on process, plot, and watershed studies conducted mostly in the second half of the twentieth century.

HYDROLOGIC RESPONSES WITHIN FORESTS

Forest disturbances and management affect the pathways of water within the forest system. Interception, evapotranspiration, infiltration, and overland (or surface) flow respond to forest disturbance and management (Figure 3-1). In turn, these changes affect watershed outputs.

Interception and Evapotranspiration

Interception is the net loss of precipitation, by evaporation, between the top of the forest canopy and the forest floor; this water is returned to the atmosphere and does not enter the soil. When forest canopies temporarily capture raindrops or suspend ice and snow, they slow the rate at which precipitation arrives at the forest floor. If this captured moisture evaporates, it effectively decreases the amount of precipitation available for soil moisture storage, transpiration, or runoff. In dispersing raindrops or suspending ice or snow, interception slows the rate at which precipitation hits the forest floor and, in doing so, effectively decreases the net effect of precipitation. Removal of trees reduces leaf area and hence, interception. Reductions in leaf area—from fire, harvest, insects, or disease—and differences in leaf area among different forest types and ages all affect hydrology in the same way (Verry, 1976; Schmid et al., 1991): a reduction in interception increases the amount of water that reaches the mineral soil. If infiltration rates are not changed, an increase in net precipitation increases soil moisture, water availability to plants, and the proportion of precipitation that is available for streamflow (Helvey and Patrick, 1965; Helvey, 1971). Reduced leaf area decreases interception rates in both rain- and snow-dominated systems; in snow-dominated systems an increase in net precipitation increases water stored in the snowpack (Neary and Ffolliott, 2005; Woods et al., 2006). Where forest canopies capture additional moisture from clouds, a reduction in leaf area can decrease net precipitation (Harr, 1982; Hutley et al., 1997; Reid and Lewis, 2007).

A reduction in leaf area also increases the amount of light reaching the forest floor, increasing energy exchange between soil or snow and the atmosphere and altering the energy budget (Figure 1-3). Increased exposure of the snowpack to solar radiation and to turbulent heat transfer by wind increases snowmelt rates relative to undisturbed forest canopies. In snow-dominated forest systems a reduction in leaf area can lead to increased snow accumulation as well as an earlier onset of snowmelt and faster melt rates (Helvey, 1980; Megahan, 1983; Hornbeck et al., 1997; Jones and Post, 2004).

The process of transferring moisture from the earth to the atmosphere by evaporation of water and transpiration from plants is called evapotranspiration. In North American forests, evapotranspiration accounts for 40 to more than 85 percent of gross precipitation. A reduction in leaf area from forest harvest, fire, or insect or disease outbreaks reduces evapotranspiration and increases water available for runoff. The magnitude and persistence of the reduction in transpiration depends on the amount and type of the vegetative canopy removed and the rate at which the vegetative cover is reestablished. However, it has only recently become possible to accurately measure transpiration in trees, and few studies have quantified transpiration rates for forest stands (but see Ryan et al., 2000; Moore et al., 2004).

Infiltration and Overland Flow

Most forests have an organic surface layer that protects the soil surface and facilitates infiltration. In most cases this water moves by subsurface pathways to the stream. Because forest soils have high infiltration rates, water rarely flows over the ground surface as infiltration excess (also called Horton overland flow). In flatter, low-lying or convergent zones, the saturated zone may rise to the surface and produce saturated overland flow.

Forest management activities and forest disturbances may remove or alter the surface layers of forest soils, and thereby reduce infiltration and increase Horton overland flow (Figure 1-1). Forest management activities and disturbances also create impervious surfaces (e.g., as roads) and modify hillslopes in ways that alter water flowpaths in soils and subsoils, shift subsurface flow to surface flow, and increase runoff and erosion rates. When organic surface layers are removed or burned, underlying mineral soil is exposed to raindrop splash and fine soil particles can accumulate on the surface, reducing infiltration and increasing overland flow. If soils are compacted to the extent that infiltration rates are lower than rainfall or snowmelt rates, the resulting overland flow can greatly increase runoff rates and surface erosion.

CHANGES IN WATERSHED OUTPUTS

Forest hydrology science describes direct changes in watershed outputs resulting from fire, timber harvest, and roads and trails (Table 3-2). These findings are summarized below.

Fire

Fire, Infiltration, and Overland Flow

Burning can greatly reduce infiltration rates and thereby increase surface runoff and erosion rates through several mechanisms: development of a water-repellent ("hydrophobic") layer at or near the soil surface; exposure of the soil surface to raindrop impact and soil sealing; increased soil erodibility; and decreased surface roughness (Box 3-2). Large post-fire increases in runoff and erosion are often attributed to an increase in soil water repellency after burning (Box 3-2), but soil sealing may play an equal or even larger role in increased runoff and erosion in some areas.

TABLE 3-2 Magnitude and Duration of Direct Effects on Watershed Outputs of Three Sets of Processes That Modify Hydrology in Forests: Fire, Forest Harvest and Silviculture, and Roads and Trails

	Processes That Modify Hydrology in Forests		
Watershed Output	Fire	Forest Harvest and Silviculture	Roads and Trails
Water yield	High-severity fire increases annual water yields; little effect of low-severity fire	Increase water yield; magnitude and duration of response varies (see text)	Little or no effect
Peak flows	High-severity fire increases peak flows; effect is short-lived	Increase peak flows; magnitude and duration of response varies (see text)	Increase peak flows; effects may be long-lived and affect extreme events
Low flows	High-severity fire increases low flows; little effect of low-severity fire	Increase low flows in short term; deficits may develop as forests regrow	Little or no effect
Erosion, landslides, sedimentation	High-severity fire increases erosion and sedimentation in streams; less effect from low-severity and prescribed fire	Increase surface erosion, landslides, and sedimentation; effects may be long-lived	Increase surface erosion (road surfaces and gullies below culverts) and landslides; increase sedimentation in streams
Water temperature and chemistry	Increases water temperature due to riparian forest removal; fire retardants and ash affect chemistry; effects are short-lived	Increase water temperature due to riparian forest removal; effects of fertilizer mostly small and short-lived; short-lived post-harvest increases in nitrate	Deliver road chemicals (e.g. salt, oil) to streams
Research gaps	Uncertainty about effects beyond a few years; magnitude and persistence of downstream effects; effects of salvage logging	Uncertainty about effects beyond one or two decades; magnitude and persistence of downstream effects; effects on habitat and aquatic ecosystems	Uncertainty about road effects on extreme floods and in watersheds >1 km^2

NOTE: These are general effects, not predictions, so qualifying adjectives such as "may," "usually," etc., are omitted. See text for factors that influence when, where, and to what extent these effects apply.

> **BOX 3-2**
> **Physical and Chemical Causes of Water Repellency in Soils**
>
> Many soils are water repellent without being burned, particularly in coniferous forests and xeric shrublands. Waxy and other aromatic compounds in the foliage of these vegetation types leach out and accumulate on the soil surface. Fungal hyphae also can generate very strong, localized soil water repellency near soil surfaces. In the absence of burning, these compounds are rarely sufficient to reduce infiltration rates at the hillslope scale. However, when burned at roughly 175-200°C, these compounds vaporize and are driven by steep heat gradients down into the soil, where they condense on cooler underlying soil particles. Thus, burning can create a semicontinuous or continuous water repellent layer at or beneath the soil surface, whose depth and thickness depend on the duration and magnitude of soil heating (DeBano, 2000; Letey, 2001). Temperatures above 280-400°C consume most waxy and aromatic compounds, so very hot fires produce a nonrepellent, disaggregated soil layer above a water-repellent layer (DeBano, 2000; Doerr et al., 2006). Coarse-textured soils are more susceptible to the formation of a water-repellent layer than fine-textured soils because of their lower surface area and greater air permeability (Huffman et al., 2001; DeBano et al., 2005).

Fire-induced soil water repellency has been well documented for certain vegetation types, particularly coniferous forests and chaparral-type ecosystems (e.g., DeBano, 2000). Fire-induced soil water repellency is spatially heterogeneous (Woods et al., 2007) and can persist for a few weeks or several years (Shakesby and Doerr, 2006). Snowmelt or prolonged rainfall may overcome water repellency until soils dry out (Doerr and Thomas, 2000; MacDonald and Huffman, 2004). Thus, burning may have less effect on infiltration and runoff during winter wet seasons or in snowmelt-dominated areas than in drier areas subjected to summer thunderstorms. Fire-induced soil water repellency breaks down by a combination of physical, chemical, and biological processes over time as plant regrowth provides a protective cover of vegetation and litter (e.g., Robichaud and Brown, 1999; Benavides-Solorio and MacDonald, 2005). Therefore, runoff and erosion rates usually return to reference or pre-fire levels within one to four years (Shakesby and Doerr, 2006).

Soil sealing refers to a reduction in infiltration as a result of breakdown of soil aggregates and rearrangement of soil particles at the surface. After moderate- and high-severity fires, rain splash can detach soil particles and reduce infiltration rates by a sealing effect at the soil surface. Combustion and the loss of organic matter also lead to a loss of soil cohesion. These effects contribute to greater overland flow (Neary et al., 1999; DeBano et al., 2005; Moody et al., 2005) and soil erodibility (Cerda, 1998; Larsen and MacDonald, 2007; Woods and Balfour, 2007).

Severe fire leads to greater reductions in infiltration and greater increases in overland flow than moderate- or low-severity fire (Shakesby and Doerr, 2006). In the Colorado Front Range, for example, summer thunderstorms with 60 mm of rainfall per hour often produce no surface runoff or erosion, but after a high-severity fire, surface rainfall intensities of only 10 mm per hour can generate overland flow (Moody and Martin, 2001; Kunze and Stednick, 2006; Wagenbrenner et al., 2006). Compared to high-severity fires, low-severity wildfires and most prescribed fires result in little or no exposure of the mineral soil surface, smaller changes in soil water repellency (Robichaud, 2000), and little effect on overland flow (Van Lear and Danielovich, 1988; Robichaud, 2000; Benavides-Solorio and MacDonald, 2005).

Salvage logging is often conducted post-fire, and although its effects on forest ecosystems are being debated (Donato et al., 2006), few studies have examined the hydrologic effects. Ground-based salvage logging generally results in more ground disturbance and less ground cover (Klock, 1975; McIver and McNeil, 2006), and extensive ground disturbance can increase soil erodibility and erosion rates. Non-ground based activities, such as the use of helicopters in logging, result in relatively little ground disturbance and may have minimal effect on post-fire runoff and erosion rates. Standard salvage logging practices are unlikely to significantly reduce or break up the water-repellent layer from a high-severity wildfire.

Fire and Water Yield

High-severity fires occur in unpredictable locations, at unpredictable times (Carpenter, 1998), which makes their study difficult and has resulted in few studies that confirm relationships between fire and water yield. By analogy to clear-cutting, high-severity fires are expected to increase water yield. Changes in interception, transpiration, and runoff processes usually lead to higher annual water yields after a high-severity fire, although the increases are highly variable between and within ecoregions (Berndt, 1971; Campbell et al., 1977; Helvey, 1980; Neary and Ffolliott, 2005; Neary et al., 2005b). Low-severity fires generally do not consume or kill enough vegetation to alter water yields significantly (e.g., Douglass and Van Lear, 1983; Gottfried and DeBano, 1990).

Fire and Peak Flows

High-severity fires can increase peak flows by one or two orders of magnitude (Scott, 1993; Moody and Martin, 2001; Neary et al., 2005b). The greatest increases in peak flows occur in areas with summer thunderstorms or fall rains, where burning has altered infiltration and overland flow processes. Reported changes in peak flows after wildfires include (1) minimal change following a high-severity burn in a snow-dominated Wyoming fir forest (Troendle and Bevenger, 1996): increase of 1.4 times in a Douglas fir forest in Oregon (Anderson, 1974); (2) increase of 6.5 to 870 times in California chaparral (Hoyt and Troxell, 1934; Sinclair and Hamilton, 1955; Krammes and Rice, 1963); and (3) increase of 20 to more than 2,000 times in ponderosa pine forests in Arizona and New Mexico (Campbell et al., 1977; Bolin and Ward, 1987; Ffolliott and Neary, 2003).

Fire and Erosion

The most important effects of fire are increases in overland flow and erosion and resulting effects on flooding and water quality. Fire can enormously increase surface erosion. Fire exposes the mineral soil and increases surface erosion; it also may increase soil moisture and landslides (Wells et al., 1979; Moody et al., 2005; Neary et al., 2005; Shakesby and Doerr, 2006). Fire-induced higher rates of erosion increase sediment delivery to streams (Helvey, 1980; Ewing, 1996; Moody and Martin, 2001; Ffolliott and Neary, 2003; Wondzell and King, 2003; Libohova, 2004; Kunze and Stednick, 2006). Low-severity and prescribed fires produce smaller effects on erosion and sedimentation in streams (Douglas and Van Lear, 1983; Van Lear et al., 1985; Van Lear and Danielovitch, 1988; Gottfried and DeBano, 1990; Wright et al., 1982). Post-fire salvage logging can further increase erosion and sediment delivery to streams (McIver and McNeil, 2006). Studies of erosion and sedimentation after high-severity fires have become more frequent in the past few decades as wildfire activity has increased on forestland in the western United States (Westerling et al., 2006).

Fire Effects on Water Temperature and Chemistry

Fire can affect a series of water quality parameters (see recent summaries by Landsberg and Tiedeman, 2000; Ranalli, 2004; Neary et al., 2005b). The effects of fire depend in large part on the pre-fire composition of organic matter and the fire intensity. The chemistry of unburned organic matter varies with plant species, underlying geology, time elapsed since the last disturbance, and atmospheric deposition of elements such as mercury and lead.

Fires usually affect water quality by the indirect pathway of increasing stream water temperature and two direct pathways, atmospheric deposition and surface runoff. Extensive burning of the riparian forest canopy removes shade, increases the amount of solar radiation, and raises stream water temperatures. Increased organic carbon and temperature in streams can reduce concentrations of dissolved oxygen (Neary et al., 2005c).

During a fire, gases and particulate matter are carried aloft and transported for varying distances before being deposited on water surfaces. In the Yellowstone fires of 1988, for example, increases in nitrogen in lakes and rivers were attributed to the diffusion of smoke into the water bodies under active fire conditions (Spencer et al., 2003).

Ash deposition can increase the pH of surface water and soil (Neary et al., 2005b). Post-fire pH values in stream water rarely exceed U.S. Environmental Protection Agency (EPA) standards (Landsberg and Tiedemann, 2000), but transient pH values of 9.5 were measured in streams after a fire in eastern Washington (Tiedemann, 1973; Tiedemann et al., 1979). Fire can cause a short-term increase in stream nitrate concentrations, and the delivery of ash and fine sediment can increase phosphorus concentrations in streams. In most cases these increases do not exceed standards for drinking water (Neary et al., 2005c).

During forest fires, chemical fire retardants are applied aerially to forests and inadvertently (perhaps unavoidably) to streams and rivers. The effects of these chemicals on water quality may be important, especially since recent studies have shown that they persist for years after application (Morgenstern, 2006). Fire retardants can contain nitrate and possibly sulfate, phosphate, and some trace elements (Landsberg and Tiedemann, 2000), which can contribute to eutrophication, especially when applied directly to streams. When these materials enter rivers, streams, and lakes, they react with sunlight to form compounds that are toxic to aquatic organisms (e.g., Buhl and Hamilton, 1998, 2000). Increased concentrations of other chemicals, such as manganese, sulfate, and mercury, also have been documented after forest fires. Elevated concentrations of both lead and mercury were detected in the post-fire runoff from the Bobcat fire outside Fort Collins, Colorado. Elevated post-fire concentrations of manganese and other constituents forced the Denver Water Board to initiate additional specialized treatments to maintain drinking water quality. In most cases the adverse effects of forest fires on chemical water quality persist for no more than two or three years.

Forest Harvest

Timber Management, Silviculture, and Water Yield

Dozens of paired watershed forest harvest experiments have demonstrated that forest removal increases water yield (Bosch and Hewlett, 1982; Hornbeck et al., 1993; Ice and Stednick, 2004; Jones and Post, 2004; Brown et al., 2005). The magnitude of water yield increases can be expressed as an absolute increase (e.g., millimeters of water, Figure 3-2) or as a percentage. Water yield increases are highest after 100 percent forest harvest, and are almost always highest in the first year after forest harvest, or the wettest year in the early post-harvest period, when changes in interception and transpiration have the greatest effect on the water balance (Bosch and Hewlett, 1982; Sahin and Hall, 1996).

Water yield increases after forest harvest vary according to several factors:

- **Climate**. The largest absolute water yield increases have occurred after cutting of forests in climates with relatively abundant precipitation (1,500 to 2,500 mm per year) and relatively low evapotranspiration. High water yield increases (300 to 500 mm per year) have been measured in the Pacific Northwest (H.J. Andrews Experimental Forest, site 1), the Northeast (Hubbard Brook Experimental Forest, site 14), and the Southeast (Coweeta Experimental Forest, site 18) (Figure 3-2). Much smaller water yield increases have been measured in regions where mean annual precipitation is low (<500 mm per year) and

(a)

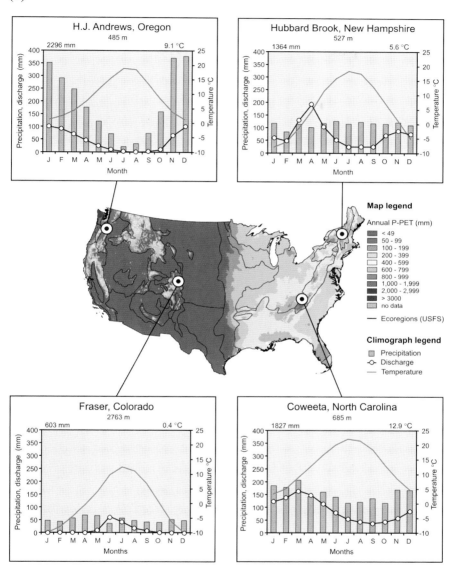

(b)

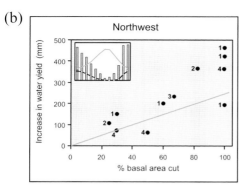
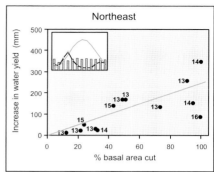
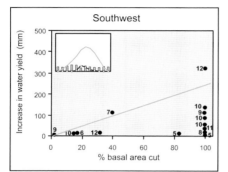
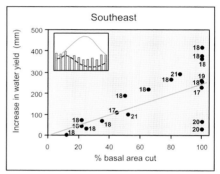

(c)

Northwest
1. Andrews, OR
2. Alsea, OR
3. Casper Creek, CA
4. Coyote Cr, OR

Southwest
5. Beaver Creek, UT
6. Castle Creek, UZ
7. Fraser, CO
8. Natural Drainage, AZ
9. San Dimas, CA
10. Three Bar, AZ
11. Wagon Wheel Gap, CO
12. Workman Creek, AZ

Northeast
13. Fernow, Wva
14. Hubbard Brook, NH
15. Leading Ridge, PA
16. Marcell, MN

Southeast
17. Alum Creek, AR
18. Coweeta, NC
19. Grant Forest, GA
20. North Mississippi, MS
21. Upper Bear Cr, Ala

FIGURE 3-2 (a) Potential annual water yield (precipitation minus potential evapotranspiration, P − PET) in millimeters for the United States, and mean monthly precipitation, temperature, and discharge at four small watershed sites in the Northwest, Northeast, Rocky Mountains, and Southeast. Mean elevation (meters), mean annual precipitation (millimeters), and mean annual temperature (degrees celsius) are shown above the graph for each of these four sites. (b) First year water yield increase (millimeters) versus percent basal area removal for 21 paired watershed forestry experiments in the conterminous United States. The trend line, shown for comparison, was taken from Figure 1 of by region were developed by Stednick (1996), which represented 95 catchments spread across similar geographic regions as shown here. (c) Locations of paired watershed sites show in (b). Black lines on map are USFS ecoregions.

potential evapotranspiration is high. In these areas, such as the interior West and Southwest, removal of forest cover is largely offset by increased soil evaporation and evapotranspiration by any remaining vegetation, and water yield increases are often <150 mm per year (Bosch and Hewlett, 1982) (Figure 3-2). Small water yield responses in the Southwest illustrate the limited potential for increasing water yields in dry forest types.

- **Seasonal timing of precipitation**. In regions where precipitation is evenly distributed throughout the year (Northeast, Southeast; Figure 3-2), water yield increases typically occur during the growing season (Martin et al., 2000; Jones and Post, 2004). In regions with dry summers and wet winters (western forests; Figure 3-2), the largest water yield increases occur in the late fall and early winter due to a reduction in transpiration and resultant increase in soil moisture carryover (Jones and Post, 2004). In snowmelt-dominated regions, most of the water yield increase occurs in spring because larger snowpacks accumulate in cutover areas (Harr et al., 1979; Troendle and King, 1985; Troendle and Reuss, 1997; Jones and Post, 2004). Thus, in both eastern and western forests, water yield increases after forest harvest often occur during seasons when water is abundant, not scarce (Harr, 1983; Troendle et al., 2001).

- **Amount of forest removed**. Forest harvest experimental treatments have ranged from 100 percent clear-cutting to partial cuts, overstory thinning, or selective harvest of a fraction of watershed area (Figure 3-2). In areas with more than 500 mm of mean annual precipitation (the Pacific Northwest, Northeast, and Southeast), water yield increases are roughly proportional to the amount of forest area cut (Hibbert, 1967; Bosch and Hewlett, 1982) (Figure 3-2). Water yield increases are difficult to detect when less than 20 percent of the basin area has been harvested (Stednick, 1996) (Figure 3-2). The spatial arrangement of cutting within a watershed also affects whether a water yield increase is detected; controlling for the amount of forest cut, there is less detectable water yield increase for thinning or selective harvests than for patch cuts (Site 4, Figure 3-2) (Perry, 2007).

- **Harvest treatments**, such as burning, herbicides, or buffer strips. Forest harvest has different effects on water yield increase depending on whether the area is burned, herbicide is applied, or the treatment is conducted in stages (compare vertical scatter of points from site 14 in Figure 3-2).

- **Storage of water in soil and snow**. Year-to-year storage of water in deep soils or poorly drained areas may offset water yield increases in areas with deep compared to shallow soils (compare site 20 versus 18, and 16 versus 14 in Figure 3-2). Post-harvest changes in snow accumulation and melt rates also can affect water yield increases after harvest (Verry et al., 1983; Troendle and Reuss, 1997; Jones and Post, 2004) (compare vertical scatter of points from site 1 or site 14 versus 13 in Figure 3-2),

- **Type and age of forest removed**. Post-harvest water yields are higher when old-growth forests with high leaf area are harvested, compared to when younger forests with low leaf area are cut (Swank and Crossley, 1988; Jones and Post, 2004). When forests of low interception (or lower annual transpiration losses), are replaced by forests with higher interception (higher transpiration losses) the net water yield can be reduced as the new forest grows to maturity (e.g., Swank and Crossley, 1988; Jones and Post, 2004).

- **Time since harvest**, or the amount of forest regrowth. As forests regenerate after harvest, water yield increases disappear. Water yield increases have persisted for as little as a decade in some areas, but for multiple decades in other areas, depending on the type and history of the forest, soils, climate, reforestation methods, and harvest treatments (Bosch and Hewlett, 1982; Troendle and King, 1985; Swank and Crossley, 1988; Hornbeck et al., 1993; Hornbeck et al., 1997; Troendle et al., 2001; Jones and Post, 2004; Brown et al., 2005). In some cases, water yields drop below pre-harvest levels after a couple of decades of forest regrowth (Hornbeck et al., 1993; Swank and Crossley, 1988; Jones and Post, 2004; Brown et al., 2005). Many paired watershed experiments established to test forest management effects on water yield were terminated after the first 5-10 years of post-treatment, so only a few paired watersheds are still providing information about the long-term consequences of past forest management for water yield.

In summary, water yield increases from forest harvest are highly variable. The highest increases in water yield occur from 100 percent forest harvest in the first years after harvest and in areas where water is relatively abundant. Because of the inherent variability in water yield responses, the amount of forest harvest necessary to produce a water yield increase varies according to regional and site-specific characteristics (e.g., amount and type of precipitation, slope, soil thickness, silvicultural methods, harvest treatments).

There is little evidence that timber harvest can produce sustained increases in water yield over large areas. Because of high evapotranspiration relative to precipitation, and dry summers, the potential for augmenting water yield on a sustainable basis in western forests and rangelands is very low (Harr, 1983; Hibbert, 1983; Troendle et al., 2001). Water yield increases from the harvest of western forests occur in winter when water is relatively abundant, and these increases would have to be stored for up to six months to effectively augment water supplies when water is scarce in late summer (Harr, 1983). Maintaining water yield increases requires continued forest harvest or conversion of forests to other land uses such as pastures, annual crops, and urban areas. Although the potential for augmenting water yield is higher in eastern than western forests, achieving this potential would require major changes in forest management objectives and land use (Douglass, 1983).

Timber Management, Silviculture, and Low Flows

Relative to peak flows or annual water yields, few studies have examined the effects of forest harvest on low flows. Most studies show an initial increase in low flows immediately after forest harvest (Harr et al., 1979, 1982; Keppeler and Ziemer, 1990; Hicks et al., 1991; Hornbeck et al., 1997; Johnson, 1998; Swank et al., 2001; Jones and Post, 2004). Observed water increases in low flows after harvest change are often short-lived, usually persisting for less than 10 years due to the relatively rapid recovery of leaf area, interception capacity, and transpiration rates.

These short-term surpluses during the low-flow period change to deficits as forests regrow. As in the case of annual water yields, the increase in low flows often is followed by a decrease in low flows to below pre-harvest levels (Hicks et al., 1991; Hornbeck et al., 1997; Swank et al., 2001). These decreases occur when a forest with relatively high transpiration and/or interception replaces a forest with relatively low transpiration or interception, such as during (1) species conversion (e.g., deciduous to evergreen) (Swank et al.,1988); (2) regeneration of a young stand with higher water use than the mature stand it replaces (Hicks et al., 1991; Perry, 2007); or (3) establishment of different riparian vegetation with greatest water demands (Moore et al., 2004; Ice and Stednick, 2004). Because relatively few studies have examined long-term trends in low flows, there is much uncertainty about this subject.

Timber Management, Silviculture, and Peak Flows

Decreases in transpiration and interception after forest harvest increase soil moisture, and higher initial soil moisture at the beginning of a storm increases storm runoff (peak flow) (Box 3-3). Recent compilations of studies examining forest management effects of peak flows show wide variability in the magnitude of peak flow response to forest harvest (Austin, 1999; Moore and Wondzell, 2005; Grant et al., 2008). Much of this variation is explained by the differences in how peak flows were defined and analyzed, dominant hydrologic regimes (e.g., rain or snow), differences in forest management, and other differences in site conditions (Austin, 1999).

Peak flow responses to forest harvest vary according to several factors:

- **Event size**. Often, the percentage increase in peak flows after forest harvest decreases as the magnitude of the peak flow increases (Harr, 1976; Beschta et al., 2000; Grant et al., 2008). However, in many cases the absolute increase in peak flows is larger with larger storms (Box 3-4; Verry, 1986; Jones and Grant, 1996; Jones, 2000; Moore and Wondzell, 2005). As storm magnitude (the total amount of rainfall or snowmelt) increases, the proportion of precipitation that can be stored by vegetation decreases. Therefore, large peak flows often experience smaller relative increases than small peak flows. Nevertheless, small percentage increases in very large floods as a result of forest harvest (Figure 3-3) may be quite large in absolute terms; a 10 percent increase in a typical 50-year flood is the same amount of water as a 50 percent increase in a 1-yr flood. Small increases in extreme floods affect more people and may be of greater concern for managers than increases in small floods.

BOX 3-3
Does Timber Harvesting Cause Floods?

The effect of forest management on flooding has been a recurrent scientific, social, and political theme (Eisenbies et al., 2007). The notion that deforestation leads to widespread land degradation and exacerbates the risk of flooding dates to antiquity (Hillel, 1994), and the role of forest management on extreme floods is an important concern for policy makers and the public (Figure 3-3). There is little doubt that forests influence the storage and movement of water, particularly at annual and seasonal time scales. Understanding the role of forest management in moderate and large floods requires a clear definition of terms and careful consideration of the various processes by which forest management can affect the size of peak flows.

Floods are variously defined by scientists and affected populations. Floods are commonly described as flows that exceed channel capacity and result in overbank inundation (Brooks et al., 2003), which can occur as frequently as every one to two years, since channels tend to adjust their shape to accommodate more frequently occurring events (Leopold et al., 1995). Hydrologists typically define floods according to their probabilities of recurrence or return period (*e.g., the 5-year flood, the 100-year flood*). Public concerns about floods are commonly limited to the more extreme events that result in a loss of life or property.

The largest floods are associated with extreme storm events, such as tropical cyclones. Some recent assessments have attempted to link the growth in flood damage in recent decades to development in flood-prone lands and to discount the role of deforestation on large-scale extreme floods (FAO, CIFOR, 2005). Nevertheless, considerable public and political pressure tends to follow extreme floods, and forest protection is often an important element of debates and policy formulation (Eisenbies, 2007).

Logging suspect in Virginia floods

By Chris Kahn
The Associated Press

IACOMA, Va. — Life along Stony Creek has never been so rough for Ray Begley.

The creek jumped its banks last summer, tearing his living room from its foundation and carrying the tangled mess downstream. Another flood last month washed away the gravel mountain road that led to his ramshackle home.

"I'm done with this place," said Begley, 62, who grew up swimming and fishing along the creek. "I just don't trust it anymore."

This is not how it used to be.

Like many people living in the mountains of Virginia's extreme southwest, Begley suspects logging above his home has created more and larger floods than ever before.

With last year's floods still fresh on people's minds and this year's floods causing $55.7 million in damage, environmentalists, hunters and residents want to stop a proposed timber harvest on 700 acres in the Jefferson National Forest, about 100 miles northeast of Knoxville, Tenn.

Local activists handed Rep. Rick Boucher, D-Va., a petition with 5,000 signatures in November condemning the proposal. In response, Boucher, who represents southwestern Virginia, has asked the U.S. Forest Service to suspend the timber sale until a panel of forest professionals and environmentalists can examine the region's timber practices.

"The flooding calls everything into question," said Blain Phillips with the Southern Environmental Law Center in Charlottesville, which represents the activists.

Bill Damon, supervisor for the George Washington and Jefferson National Forests, said his agency has tried to be responsive to public concern, cutting the proposed harvest in half from 1,413 acres. Damon said he has trouble understanding why there is such resistance to logging on just a few hundred acres in a forest of 1.76 million acres.

Damon would not say if he believes logging contributes to flooding, deferring to Boucher's advisory committee, which started meeting in February.

The panel, which includes local officials, activists, the Forest Service, U.S. Fish and Wildlife, the Army Corps of Engineers and lumber industry representatives, is expected to make its recommendations by summer.

In September, however, the Forest Service concluded in a field study that July's floods on Stony Creek would have happened no matter what.

"That was a big flood," said Gary Kappesser, a Forest Service hydrologist. "Any management within the watershed would have made little difference."

Similar complaints have been raised in other parts of Appalachia, with many people suggesting that the coal mining practice of cutting away mountaintops has also made floods more dangerous.

Michael Gillen, a hydrologist with the National Weather Service in Blacksburg, said flooding has increased all along the Appalachian range, but he is not sure why. The effects, however, are increasingly visible as more people move to rural areas, he said.

"You've got more people living along these streams because they're so beautiful when they're not flooded," Gillen said, "and when you clear-cut property, you reduce its ability to absorb runoff."

FIGURE 3-3 Public perception of flooding is often linked to land use activities. SOURCE: Reprinted, with permission, from Associated Press (2002). Copyright 2002 by Associated Press.

> BOX 3-4
> Regional Variability in Peak Flow Response: Northern Lake States
>
> The effect of timber harvesting (or forest clearing for other uses) on the timing of peak flows varies regionally, as illustrated in the northern Lake States, particularly where snowmelt dominates the annual hydrograph. In mountainous areas, snowmelt is naturally desynchronized by the heterogeneity of terrain features such as slope, aspect, and elevation. The snowpack generally melts first at low elevations and last at high elevations, more quickly on south slopes than on east and west slopes, and most slowly on north slopes. Differences in forest canopy type and density (e.g., dense, even-aged conifers versus sparse, deciduous stands) also change the timing and rate of snowmelt. In the relatively flat terrain of the northern Lake States the effect of topography on energy exchange is minor in comparison to the influence of forest cover. Therefore, when timber harvesting (e.g., patch or strip cuts) changes the energy balance and microclimate of a watershed, it tends to desynchronize snowmelt (Figure 3-4). In effect, forest harvest changes the snowmelt hydrograph from one peak to two, with the first peak coming from the harvested openings and the second peak from the remaining mature forest. As the proportion of the harvested area within a watershed increases (e.g., greater than ~70 percent open), snowmelt occurs earlier and at a more rapid rate.
>
>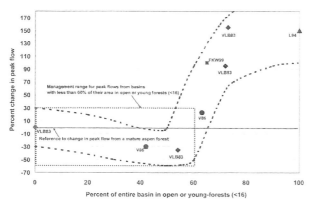
>
> NOTE: Dashed lines are theoretical envelope curves started at ±30 percent on the left but expanding to include all of the data to the right. They are an estimated mean response first proposed in Verry (1986). Data points (author and year of publication) are from paired watershed experiments (Verry, Lewis, and Brooks 1983 — Marcell Experimental Forest), double mass curve historical evaluations (Verry 1986 — Upper Mississippi above St. Paul, Minnesota), historical geomorphic and modeling analyses (Fitzpatrick, Knox, and Whitman 1999 — North Fish Creek near Ashland, Wisconsin), and modeling of complex upland/peatland watersheds (Lu 1994 — Marcel Experimental Forest). Several state, county, and river basin groups use the 60 percent open and young forest condition to guide watershed cover condition.
>
> FIGURE 3-4 First-year increase in peak discharge versus percent forest clearing for watersheds in the northern Lake States. SOURCE: Reprinted, with permission, from Ice and Stednick (2004). Copyright 2004 by the Society of American Foresters.

- **Rain versus snow**. Storm events involving rain respond differently to forest harvest than those involving snow. In rain events, forest harvest affects peak flows directly through changes in soil water. In events involving snow, the effect of forest harvest on peak flows depends on how forest harvest changes snowpack size and snowmelt, as well as soil moisture (Verry et al., 1983; Troendle and King, 1985; Jones, 2000; MacDonald et al., 2003; MacDonald and Stednick, 2003; Schnorbus and Alila, 2004).
- **Season**. Peak flow increases after forest harvest are proportionately larger in spring, summer, and fall compared to winter, because soil moisture levels are sensitive to transpiration by forest vegetation in spring, summer, and fall. However, peak flow increases (in millimeters) after forest harvest are absolutely (see Figure 3-2) larger in winter than in spring, summer, or fall, because peak flows are higher in winter, and reductions in transpiration in previous seasons carry over into winter (Jones, 2000; MacDonald and Stednick, 2003).
- **Proportion of area harvested**. The larger the proportion of area harvested, the greater is the increase in peak flows (Jones, 2000; Moore and Wondzell, 2005). Peak flow increases have been detected after only 25 percent harvest of a small watershed (Harr et al., 1979, 1988; Jones and Grant, 1996; Caissie et al., 2002).
- **Topographic relief and elevation**. The effect of forest harvest on the energy balance, and resulting changes in snow accumulation and melt, vary with elevation and aspect. Forest harvest may increase peak flows during rain-on-snow events in the Pacific Northwest (Harr, 1981; Harr, 1986). In the flatter topography of the northern Lake States, harvesting 20 – 50 percent of the watershed desynchronizes snowmelt and reduces annual snowmelt peak flows by as much as 40 percent, while harvesting over 60 percent of a basin can increase the size of snowmelt peak flows by more than 140 percent (Box 3-4 and Figure 3-4) (Verry, 1986).
- **Time since harvest**. As forests regenerate, peak flows return to pre-harvest levels (Troendle and King, 1985; Jones, 2000).
- **Roads and skid trails**. Many studies of forest harvest effects on peak flows include some roads and skid trails, which can accentuate the effect of harvest on peak flows (Jones and Grant, 1996). Because roads and trails influence different components of the water balance, they are discussed separately below.

Timber Management, Silviculture, and Erosion, Mass Movement, and Sedimentation

Many studies have shown that timber harvest practices greatly increase surface erosion (summarized in Dunne and Leopold, 1978; Brooks et al., 2003). Overland flow and surface erosion are rare in undisturbed forests, but logging operations expose surface soils and lead to surface erosion. Multiple studies have shown that surface erosion is most significant in areas of soil disturbed by cable yarding and skidding of cut logs to landings (e.g., Johnson and Beschta,

1980). Many forestry regulations govern surface erosion and sediment production. The effects of skid trails and unpaved roads on surface erosion are described below in the discussion of roads and trails.

In steep landscapes, extreme storms trigger landslide events and the associated input and transport of bedload and woody debris (MacDonald and Coe, 2007). Portions of the Pacific Northwest, northern Rocky Mountains, and central Appalachians are especially prone to shallow landslides. After forest harvest on steep slopes, decreasing root strength and increased soil moisture and pore water pressures contribute to decreased slope stability and can increase the likelihood of shallow landslides (debris avalanches) during precipitation events. Higher soil moisture (from reduced interception and transpiration) increases the forces generating slope movement. Higher soil moisture also increases pore pressures and reduces soil cohesion; combined with loss of root strength after harvesting, these factors reduce the forces resisting slope movement (Swanson and Dyrness, 1975; Sidle et al., 1985; Montgomery et al., 2000; Miller and Burnett, 2007). A number of studies have documented increased landslides from forest harvest relative to undisturbed forested areas (Swanson and Dyrness, 1975; Sidle and Ochiai, 2006). Forest clearcutting may increase the landslide erosion rate by two to nine times relative to undisturbed areas (Montgomery et al., 2000; Sidle and Ochiai, 2006; Miller and Burnett, 2007). Sediment and woody debris delivered to stream channels by landslides can be transported downstream by debris flows, which may create debris dams and exacerbate flooding in lowland settings, causing extensive property damage. Steep forestlands also are prone to deep slow-moving earthflows. During large storms, these earthflows may contribute material to streams, including fine clays, which may create persistent turbidity in downstream reservoirs and water supply systems (Bates et al., 1998).

Many studies have shown that fine sediment contributed to streams by surface erosion from exposed soils or landslides can greatly increase suspended sediment in streams, adversely affecting aquatic habitat, especially in steep, coarse-bedded streams (e.g., Campbell and Doeg, 1989). Suspended sediment levels may remain elevated for many years or decades after timber harvest (Grant and Wolff, 1991).

Timber Management, Silviculture, Water Temperature, and Chemistry

Removal of riparian vegetation along streams causes peak water temperatures to increase as a result of increased solar radiation. The largest stream temperature increases occur in the summer (Levno and Rothacher 1967, 1969; Beschta and Taylor, 1988; Binkley and Brown, 1993a; Johnson and Jones, 2000). As streamside forests regenerate and provide shade, temperatures usually return to pretreatment levels (Johnson and Jones, 2000; Ice et al., 2004). Maintaining streamside forests of sufficient width to shade streams helps mitigate temperature increases from forest harvest (Stednick, 2000; MacDonald and Coe, 2007), but groundwater contributions and hyporheic flow also mitigate stream temperature increases (Story et al., 2003; Johnson, 2004).

In silviculture, fertilizers, herbicides, fungicides, insecticides, and fire retardants are used to protect and enhance tree growth. Plant available nitrogen and phosphorus are occasionally added to forests as fertilizer if one or both of these nutrients are limiting growth. Generally, increases in stream concentrations after fertilization are minimal because forest soils efficiently retain nutrients, and because fertilizer that is not absorbed is often volatilized (Binkley et al., 1999; NCASI, 1999; Stednick, 2000). If fertilization increases stream nitrogen concentrations, the maximum value is usually reached within two to four days, and concentrations decrease rapidly thereafter (although a return to pretreatment conditions could take six to eight weeks). Greater effects on stream nutrient concentrations occur when fertilizer is applied directly to the stream, applied during rainy periods (Stednick, 2000), or applied to sites already affected by nitrogen pollution from non-fertilizer sources (Fenn et al., 1998). Relatively few studies have addressed the effects of fertilizer at large watershed scales or over the long term (Anderson 2002; McBroom et al., 2008).

When pesticides are applied according forestry regulations, they are unlikely to impair water quality (Michael, 2000; Dent and Robben, 2000 [from Ice et al., 2004]; Tatum, 2003, 2004). If pesticides enter streams, they are usually at low concentrations and only remain for a short period of time (Ice et al., 2004). Nevertheless, members of the public, nongovernmental organizations (NGOs), and the scientific community remain concerned that risk assessments are incomplete and that manufacturers funded most studies. Concerns focus on (1) chemicals that have yet to be investigated; (2) synergistic, cumulative effects of mixtures in the environment; (3) effects of degradation products in the long term through transformation and transport in groundwater (Michael, 2000); and (4) inadequate tests of how stream ecosystems might react because native amphibians may be more sensitive than laboratory animals (Tatum, 2003).

Most forests are naturally nitrogen-limited, and stream concentrations of nitrogen are lowest in young forests and increase as forests mature (Edwards and Helvey, 1991; Swank and Vose, 1997; Vitousek, 1997). Decades of elevated atmospheric deposition of nitrogen (and sulfur, in the eastern United States) have greatly altered the nitrogen dynamics of eastern forests (Likens et al., 1977, 1996; Lovett and Kinsman, 1990; Likens and Bormann, 1995; Fenn et al., 1998) and increasingly, western forests (Riggan et al., 1985; Binkley, 2001). Resulting nitrogen saturation of soils and streams (Aber, 1992), as well as soil acidification and loss of basic cations, has affected forest growth and tolerance to cold stress, elevated stream nitrogen concentrations, and acidified streams, especially in the eastern United States (Adams et al., 1993; Lovett and Lindberg, 1993; Flum and Nodvin, 1995; Rustad et al., 1996; Lawrence et al., 1997; De Hayes et al., 1999; Lawrence and Huntington, 1999; Lovett et al., 2000; Baldigo and Lawrence, 2001). Riparian forest buffers have been adopted to mitigate these effects (Lowrance et al., 1997).

After forest harvest, concentrations of nitrate-nitrogen (nitrate-N), one of the most mobile nutrients in disturbed forests, typically increase in streams as soil moisture content and subsurface flow rates increase (Likens et al., 1970; Miller and Newton, 1983; Martin and Harr, 1989; Binkley and Brown, 1993; Martin et al., 2000). Elevated stream nitrogen concentrations return to pre-harvest levels within a few growing seasons or less as young forests grow on a harvested site (Likens et al., 1970, 1977; Martin and Harr, 1989; Likens and Bormann, 1995; Martin et al., 2000; Swank et al., 2001). The duration of elevated post-harvest nitrogen in streams varies among sites according to soil nitrogen levels and microbial activity, atmospheric deposition, and forest regrowth (Swank, 2000). A review of more than 40 studies found that nitrate-N on averaged doubled in concentration to 0.44 mg/L for one to five years after harvest on 75 percent of the locations, but the remaining studies could not detect increases and a few (5) even had 24-95 percent declines (Binkley et al., 2004).

Elevated peak flows and surface erosion after forest harvest may increase phosphorus delivery to streams. Phosphorus export occurs most during stormflow events that mobilize the fine particulate matter to which phosphorus is sorbed (Hobbie and Likens, 1973; Meyer and Likens, 1979). Riparian forest buffers have been successful in reducing particulate phosphorus (phosphorus attached to sediment) by minimizing overland flow but less effective in removing dissolved phosphorus (de la Crétaz and Barten, 2007; Stednick, 2000).

Roads

Roads, Trails, Infiltration, and Overland Flow

Roads and skid trails modify surface and subsurface flowpaths of water (Figure 3-5). Forest roads alter runoff processes in two ways (Figure 3-5) (Megahan, 1972; Wemple and Jones 2003). First, roads and skid trails (compacted soil surfaces) generate overland flow because they have very low infiltration rates (Johnson and Beschta, 1980; Luce and Cundy, 1994; Ziegler and Giambelluca, 1997). Cutslopes above roads and hillslopes below roads also often have lower infiltration rates than forest soils, and they also may generate overland flow. Roads constructed on steep slopes also can intercept water flowing in the subsurface, further increasing overland flow (Megahan, 1972; Wemple and Jones, 2003). During precipitation or snowmelt events, water flows on road surfaces or in ditches that are connected to streams, so the road network delivers this water directly to the stream network.

Roads also increase overland flow by intercepting water flowing in subsurface flowpaths at cutbanks on steep hillslopes (Figure 3-5) and converting this water to surface flow, which the road network then delivers to streams (Wemple

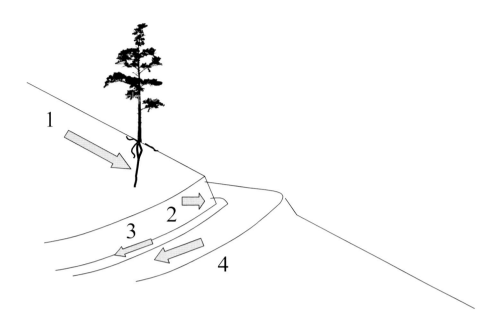

FIGURE 3-5 Schematic diagram of how roads modify water flowpaths in soils and subsoils. Roads generate Horton overland flow and intercept subsurface flow: 1 = shallow subsurface flow in the hillslope; 2 = subsurface flow intercepted by the roadcut; 3 = water flowing in roadside ditch; 4 = Horton overland flow on impervious road surface. Adapted from Wemple et al. (1996).

and Jones, 2003). The greater the depth of the soil profile exposed by the cutbank, the greater is the potential for subsurface flow interception (see photo below). With these two effects, roads convert relatively slow moving subsurface flow to overland flow, which has higher flow velocities.

Roads, Trails, and Water Yield

Roads have their greatest impact on hydrologic effects on infiltration and overland flow (see preceding section). Neither in empirical studies nor in modeling studies have roads been identified as having significant effects on annual water yields in forested systems.

An unpaved forest road in steep terrain with a large cutslope. Photo courtesy of B. Wemple.

Roads, Trails, and Peak Flows

Roads redistribute water locally and alter flow routing. Roads can contribute to an increase in the size of peak flows by increasing the amount of surface runoff from impervious surfaces, intercepting subsurface stormflow, and speeding the delivery of this runoff to the stream network through ditches or gullies (Megahan, 1972; Wemple et al., 1996; Wemple and Jones, 2003). The percentage of unpaved roads that are connected to the stream network is directly proportional to mean annual precipitation and decreases with the presence of engineered road drainage structures such as waterbars, rolling dips, and relief culverts (Coe, 2006). Through the combination of increasing the amount of surface runoff and delivering this runoff more rapidly to the stream channel, roads can produce detectable changes in peak flows at the small watershed scale (Harr et al., 1975; Ziemer, 1981; King and Tennyson, 1984; Wright et al., 1990, LaMarche and Lettenmeier, 2001). Modeling studies have replicated measured road effects on peak flows (Bowling and Lettenmeier, 2001; LaMarche and Lettenmeier, 2001; Tague and Band, 2001; Cuo et al., 2006).

It is debated how much roads affect very large peak flows in small or large watersheds (Jones and Grant, 1996; Megahan and Thomas, 1998; Beschta et al., 2000; Jones, 2000; Jones and Grant, 2001a, 2001b). A small number of studies designed to test road effects and a lack of long-term records that capture extreme floods contribute to uncertainty about the magnitude of road effects on very large floods or in large watersheds. Very few experimental studies have been conducted with road-only treatments, but many paired watershed forest harvest experiments include roads. Very few large watersheds lack roads, and fewer of these have streamflow records. Hence peak flow responses to forest harvest often include the effects of forest removal, the effects of roads, and the interaction between them (Jones and Grant, 1996).

Roads, Trails, Erosion, Mass Movements, and Sedimentation

High rates of overland flow along unpaved road surfaces entrain sediment, erode road surfaces, and contribute fine sediment to forest streams (Reid and Dunne, 1984). Overland flow on road surfaces and in roadside ditches and culverts concentrates soil moisture on oversteepened fillslopes below roads and cutslopes above roads, increasing susceptibility to landslides (Swanson and Dyrness, 1975; Larsen and Parks, 1998; Wemple et al., 2001; May, 2002). Forest roads can increase the landslide erosion rate by 30-300 times relative to undisturbed areas (Sidle and Ochiai, 2006), much more than the effects of forest harvest. Road fills and cutslope areas are subject to landslides during storm events, and these can contribute large volumes of sediment to downstream receiving waters (Swanson and Dyrness, 1975; Wemple et al., 2001). Landslides contribute fine sediment to streams, which can be detrimental to water quality and aquatic habitat, as well as coarse sediment and large pieces of wood, which are important structural elements for stream ecosystems but also can exacerbate downstream flooding and serve as tools for damaging roads and bridges.

In addition to their effects on landslides, compacted road surfaces, cutslopes, fillslopes, ditches, and areas below culverts are exposed to chronic surface erosion as a result of the generation and concentration of overland flow; roads deliver much of this fine sediment directly to streams, where it may become suspended sediment (Reid and Dunne, 1984; Bilby 1985). Suspended sediment is the most ubiquitous nonpoint pollution source from forests (Landsberg and Tiedemann, 2000) and can degrade aquatic ecosystems by reducing water clarity, reducing interstitial flow and dissolved oxygen levels, and altering stream channel morphology (Waters, 1995; Stednick, 2000; Swanson et al., 2000). Sediment-laden water increases water treatment costs, reduces water storage facility storage volume and life span, and interferes with disinfection processes (NRC, 2000; Scatena, 2000; Stednick, 2000). Sediment particles also can bind with and become a transportation vehicle for contaminants such as nutrients, metals, organic compounds, and pesticides.

MANAGING FORESTS FOR WATER

Forest hydrology principles elucidate direct effects of forest management and disturbance on hydrologic processes. Direct effects of forest management and disturbance on hydrology include increased net precipitation, temporary increases in water yield, and increased suspended sediment concentrations. General principles of forest hydrology science indicate that increased water yield is one of the direct effects of forest harvest (Tables 3-1 and 3-2). Increases in water yield occur locally and may last up to a few decades after forest harvest.

Although in principle forest harvest can increase water yield, in practice a number of factors make it impractical to manage forests for increased water. Water yield increases from vegetation removal are often small and unsustainable, and timber harvest to augment water yield may diminish water quality. Increases in water yield tend to occur at wet, not dry, times of the year, and tend to be much smaller in relatively dry years. In addition, harvesting enough area to achieve a sustainable increase in water yield will have potential effects on wildlife fisheries and aquatic ecosystems

Forest hydrology principles also describe **indirect** and **interacting** effects of forest management on hydrologic processes. Indirect effects are responses to forest management that are displaced in time or space, such as fire suppression leading to insect outbreaks that affect forest hydrology. Interacting effects occur when two or more management practices coincide, such as when post-salvaging logging and road building have a different collective effect on forest hydrology than their individual effects. The state of knowledge of forest hydrology provides a strong foundation for knowledge of the **direct** effects of forest management and disturbance on hydrology. Contemporary forest management and disturbance processes raise issues that require extending the science from these principles to prediction, including indirect and interacting effects of changes in forest landscapes. These research needs are discussed in Chapter 4.

4
From Principles to Prediction: Research Needs for Forest Hydrology and Management

Forest hydrology has built a strong foundation of general principles (Table 3-1) concerning the direct effects of forest management on hydrologic processes (Table 3-2) from plot studies, process studies, and watershed experiments (Chapter 3). The challenge now is to apply these principles to predict how hydrologic processes will respond to many forms of change in forest landscapes.

Forest hydrologists have long recognized the need to understand indirect and interacting effects of forest management at much larger spatial scales and longer temporal scales than is possible in plot studies, process studies, and watershed experiments. Indirect effects are responses to forest management that are displaced in time or space, such as fire suppression leading to insect outbreaks that affect forest hydrology. Interacting effects occur when two or more management practices coincide, such as when post-salvage logging and road building have a different collective effect on forest hydrology that differs from their individual effects.

This chapter examines the research challenges faced by forest hydrology as it moves from principles to prediction at larger spatial scales, at longer temporal scales, and in a changing social context. The chapter concludes by outlining the potential for improved cumulative watershed effects analysis that could provide the predictions needed by forest and water managers in the twenty-first century.

SPATIAL RESEARCH NEEDS

A key unresolved issue in forest hydrology is how to apply the findings of hydrological studies in one area to a different area or how to scale up the findings to large watersheds and landscapes (defined in Chapter 1, Box 1-1). Although forest hydrologists have confidence in the general principles of hydrologic responses to forest management and disturbance (Chapter 3), they cannot predict precisely how forest management will affect hydrologic processes in specific places other than those that have been intensively studied. Most forest hydrology studies are conducted in small watersheds that are instrumented to measure streamflow and other hydrologic properties. However, the sum total of the area studied by forest hydrology is only a tiny fraction of the watersheds in the United States, and hydrologists recognize the need to extend hydrologic knowledge from "gauged basins" (watersheds that have measured records) to ungauged basins. Predictions are most needed in ungauged places to better understand hydrologic effects where conflicts sometimes arise: in water supply systems for agriculture and cities; rivers where endangered and threatened aquatic species occur; large water bodies such as the Chesapeake Bay; and many, many others.

Forest hydrology is adopting a landscape perspective to examine spatial patterns of forests and associated hydrologic processes and to link principles from plot- and small watershed scales (up to several square kilometers, see Chapter 3) to predictions at larger spatial scales (hundreds of square kilometers). Within a watershed, forests can be located in the headwaters, along riparian corridors, in woodlots in agriculture lands, and in urban or suburban areas (Figure 4-1). Based on its intra-basin position, a forest fulfills various water-related functions with respect to water quantity and quality. For example, forests in headwaters influence water yield and the quality of water delivered to downstream areas. Riparian forests located along streams throughout a watershed provide key functions for protecting streams from inputs of sediment (Naiman and Decamps, 1997), nutrients, and herbicides (Peterjohn and Correll, 1984; Lowrance et al., 1997); provide wildlife habitat for terrestrial and aquatic organisms (Barton et al., 1985; Darveau et al., 1995); and support a diversity of other functions (Risser, 1995). Riparian forests have been greatly altered by economic development, and they are the focus of many forest management guidelines designed to preserve water quantity and water quality.

Research Need: Process studies are needed to determine how forests, particularly riparian forests, affect water quantity and quality according to their position within a watershed.

Hydrologists use models to predict water quantity and quality in watersheds where there are no measured records. Since 2004, a working group for Prediction in Ungauged Basins (PUB) of the International Association of Hydrological Sciences (*http://www.hydrologic science.org/pub/about.html*) has developed methodologies for assessing uncertainty in hydrologic predictions arising from uncertainties in landscape properties and climate inputs, choice of model structure, and methods of information transfer from gauged to ungauged watersheds (Sivapalan, 2003). Most hydrological models are developed and tested for gauged basins and subsequently are validated and applied to ungauged areas. However, models that have been fitted to data in small, gauged watersheds often provide inaccurate or imprecise predictions when they are (1) extrapolated to other small forested headwater basins, (2) extrapolated to future time periods, or (3) applied to large watersheds. This problem of prediction in ungauged basins has preoccupied hydrology researchers for several decades, and is compounded by a lack of information about how direct hydrologic effects interact under the multiple sets of specific conditions that occur in changing forest landscapes. By examining forest hydrologic processes under a wide range of conditions, landscape-scale studies could provide data and understanding to help extend basic forest hydrology principles to make predictions needed by water managers.

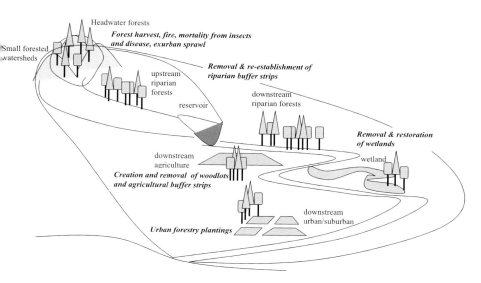

FIGURE 4-1 Changes in forests in various parts of watersheds collectively contribute to cumulative watershed effects. Bold italicized font illustrates a few of the changes occurring to forests in these parts of watersheds in the United States. Harvest, fire, insects, and pests in headwater forests alter quantity and quality of water delivered to downstream areas. Modifications of riparian forests and wetlands affect quantity, quality, and timing of water delivered from lands adjacent to streams and rivers. Modifications of forests—small woodlots, windbreaks—on agricultural lands and urban forestry can influence water quantity and quality.

Research Need: Landscape-scale studies to improve predictions of hydrologic responses in large watersheds and landscapes based on general principles of hydrologic responses to forest management (Tables 3-1 and 3-2) developed in small, homogeneous watersheds

Road networks are a pervasive feature of forest landscapes (Figure 4-2). The location and density of roads in a watershed can influence the hydrologic effects of roads. In many forested areas, roads are concentrated in the valley bottoms immediately adjacent to streams, meadows, and wetlands. These roads have a particularly high potential for delivering runoff and sediment to streams, lakes, and aquatic ecosystems. The legacy of midslope roads from past logging practices is also of concern, because these have a high potential for subsurface flow interception, connectivity to the drainage network, and initiating shallow landslides. Considerable research effort has been devoted to modifying road design and management to mitigate erosion, and to developing techniques

FIGURE 4-2 A dense network of forest roads in the Cascade Mountains of western Oregon. Photo courtesy of A. Levno, U.S. Forest Service, retired.

to decommission roads (Madej, 2001; Megahan et al., 2001; Ice et al., 2004; Switalkski et al., 2005). The hydrologic effects of road networks at large scales, and the effects of road decommissioning are not widely studied.

Research Need: Landscape-scale studies of the effects of road networks and road decommissioning on water quantity and quality in larger watersheds and landscapes, particularly during extreme storm events

TEMPORAL RESEARCH NEEDS

Water management systems have been designed and operated under the assumption that hydrologic variables such as annual water yield, while varying over time, can be predicted reliably based on instrument records. However, increased understanding of long-term variability and trends in climate has undermined this assumption (Milly et al., 2008; Barnett et al., 2008). If precipitation and streamflow vary over the long term, future annual water yields may fall short of the levels that water supply systems—and attendant agricultural and urban development—were designed to provide. Given this context, it is critical for forest hydrologists to extend beyond general principles to make predictions of how forest management and disturbance affect hydrologic response on time scales that exceed those of most forest hydrology science.

Some forest and stream management plans now include the historical range of variability, which presupposes that (1) past conditions and processes provide context and guidance for managing ecological systems today; and (2) disturbance-driven spatial and temporal variability is a vital attribute of nearly all ecological systems (Landres et al., 1999; Perera et al., 2004; Poff et al., 1997). The historical range of variability helps characterize the variation in quantity, quality, and timing of streamflow from forests. It can also be used to establish baselines for assessing change in water from forests over time.

Forests and their associated hydrologic processes change on time scales ranging from decades to hundreds, or even thousands, of years (Table 3-2). The temporal context for understanding forest hydrologic processes involves expanding the temporal scale into the past to consider the effects of past forest practices and into the future to project and anticipate changes in land use and climate. Many different kinds of legacies of past human activities affect forests in the United States (USGCRP, 2000). In some areas, native forests have been converted to agricultural and urban uses, and forests have regrown on abandoned agricultural lands in others. Roads and expansion of urban areas have fragmented forests into smaller, less-contiguous patches and created new drainage patterns. Fire suppression has changed the structure and community composition of many forests, especially in those with otherwise active fire regimes. Exotic species introductions, grazing by domestic animals, predator eradication, and timber harvesting methods have changed forest cover and species composition as well. Future urban and suburban development and climate change are expected to continue to alter forest cover and species composition. Human activities will modify forests in the future (USGCRP, 2000), and future legacies will reflect current, regional forest histories (NRC, 2002).

Research Need: Long-term predictions of hydrologic responses to forest management, including harvest, roads, and fire suppression, over decades to centuries based on general principles of hydrologic responses to forest management (Table 3-2)

Regional Forest Histories in the United States

Each region of the United States has a different history of forest conversion and management by humans that has yielded forests of different types with different capacities to produce clean, abundant water. Hydrologic effects of regional forest histories in most areas were not documented and may be difficult, but not impossible, to reconstruct. Although data collection and recording at many experimental watersheds ceased in the 1970s and 1980s, these forested properties remain in the public domain and hydrologic data collection could be reactivated at these sites. If resumed, streamflow and water quality monitoring at these sites could be very informative about the hydrologic effects of long-term changes in forests, land use and land cover. For example, research conducted at the Coweeta Hydrologic Laboratory (and the Long-Term Ecological Research [LTER] Network) shows that pine plantations, which occupy much of the forest area in the Southeast, use more water than native deciduous forests (Swank et al., 1988). These types of assessments strengthen the understanding of hydrologic effects of forest management activities.

Research Need: **Studies that compare hydrologic responses to forest management and long-term changes in forest species among the various regions of the United States.**

Legacy of Exotic Species

Many forested landscapes in the United States are affected by nonnative species introductions, a legacy of past human effects on landscapes. Introduced species cause profound ecological effects (Mack and D'Antonio, 1998), but their direct and indirect effects on hydrology are not as well documented. Possible declines in water yield resulting from the invasion of riparian zones by exotic tree species such as salt cedar (*Tamarix* spp.) and Russian olive (*Elaeagnus angustifolia*) have been a major source of concern in the arid southwestern of the United States (Vitousek, 1990). Riparian vegetation in southwestern rivers, including nonnative salt cedar and Russian olive, as well as native cottonwood, may use up to one-third of the water lost along the river, or amounts equivalent to irrigation withdrawals and evaporation (Dahm et al., 2002). Simple strategies to increase water yield in arid regions by tree removal, including exotic invasive tree species, have not produced consistent, demonstrable increases in water yield (Shafroth et al., 2005). Hydrologic processes in forests also are affected indirectly by defoliation and tree mortality due to introduced insects (e.g., gypsy moth, hemlock wooly adelgid) and diseases (chestnut blight).

Research Need: **Studies of the direct effects of exotic tree species introduction or removal and indirect effects of introduced insects and diseases, on water quantity and quality from forests.**

Legacy of Fire Suppression

The legacy of fire suppression, practiced since European colonists arrived and especially since 1910 in the western United States, may have effects on forest structure and water (Clark, 1990; Baker, 1992; Covington and Moore, 1994). One consequence of fire suppression has been an increase in stem density and leaf area in forests (Johnson et al., 2001; Graham et al., 2004). Another consequence of fire suppression has been an increased susceptibility to insect and pest outbreaks and greater vulnerability to defoliation during outbreaks, particularly in forests where the suppression of fire has led to crowding of trees and increased stress (Bergeron and Dansereau, 1993; McCollough et al., 1998; Power et al., 1999). As time since the last fire increases and forest age exceeds the natural fire return interval, outbreaks of insects such as spruce budworm (*Choristoneura fumiferana*), spruce beetle (*Dendroctonus rufipennis*), and mountain pine beetle (*Dendroctonus ponderosae*) are more likely, and mortality from outbreaks is higher. Connections between fire suppression and insect epidemics have been documented in Alberta, British Columbia, the Rocky Mountains, and the Pacific Northwest (Bergeron and Leduc, 1998; Bebi et al., 2003; Taylor and Carroll, 2004).

In small paired watersheds, forest mortality as the result of insects and pests produces a short-term increase in water yield and some transient effects on water quality (Swank, 1988; Lewis and Likens, 2007), but in the longer term, foliage regeneration may lead to decreases in water yield after insect infestations (Swank et al., 1988). However, very few studies have addressed the hydrologic consequences of fire suppression and insect outbreaks (Love, 1955; Bethalmy, 1974, 1975; Alila et al., Year). The longer-term, indirect and interacting effects of fire suppression on forest ecology, water yield, and water quality over large watersheds are difficult to estimate because the changes in forest density and composition are largely undocumented. Data exist and could be used to assess the effects of fire suppression on water yield and quality in some experimental watersheds owned and managed by federal agencies (the U.S. Forest Service [USFS] and other agencies); however, these data have not been analyzed for these purposes.

Research Need: Landscape-scale, long-term studies of the effects of fire suppression and insect and disease outbreaks on water quantity and quality from forests.

Legacy and Future of Grazing, Predator Eradication, and Predator Reintroduction

Grazing by domestic livestock and native animals and the consequences of eradication and reintrodution of predators have potential but largely unquantified effects on hydrology from forested watersheds. Grazing of forests by domestic livestock in the nineteenth and twentieth centuries had long-term consequences for forest ecology, water timing, and water quality (Rummell, 1951, Johnson, 1952; Armour et al., 1994). Forest grazing by sheep and cattle left biophysical legacies in forest landscapes. It suppressed fire, helped convert the original park-like forests of the interior western United States into dense stands of less fire-tolerant tree species, and changed the physical environment by reducing fire frequencies, compacting soils, reducing water infiltration rates, and increasing erosion (Belsky and Blumenthal, 1997; Graham et al., 2004). Continued mobility of sediment accumulated in stream channels from elevated upland erosion or historic agriculture in forests in the past (e.g., Platts, 1981) is another landscape-scale legacy of forest grazing that may still be influencing water quality.

In forests throughout the United States, eradication of native predators has led to increases in populations of deer and other browsers, reducing the cover of tree seedlings and saplings, particularly in riparian forests (Terborgh et al., 2001). Ecologists call the indirect effects of predator removal on forest vegetation "trophic cascades," whereby predators exert "top-down" control on primary production and growth of vegetation (Polis et al., 2000). In Yellowstone National Park, ecological studies indicate that wolf eradication increased browsing and largely eliminated riparian aspen forests; the reintroduction of wolves and associated trophic cascades may lead to riparian vegetation expansion in some areas (Ripple and Beschta, 2007). Despite a rapidly expanding literature on ecological trophic cascades, very little work has examined the indirect hydrologic effects on water yield and quality from trophic cascades and predator eradication and reintroductions.

Research Need: Studies of the indirect and interacting effects on water yield and quality of reduced grazing by domestic cattle and sheep and of predator removal and reintroduction on ungulate browsing in riparian forests.

Future Climate Change Effects on Forests

Climate changes over the past half-century are likely to have major effects on water quantity and timing from forests; some of these effects are already apparent. Climate change effects on forests will occur through (1) direct effects on precipitation type, snowmelt timing, and amount of precipitation; (2) indirect effects on disturbances in forests, including fire, wind, insects, and pests; and (3) indirect effects on vegetation species ranges including both natives and exotics.

A number of modeling studies have investigated direct hydrologic effects of climate change at the large watershed, regional, national, and global scales (Barnett et al., 2004). Simulated future climate in the Columbia River Basin indicate a shift in the timing of water availability through reduced snowpack, earlier snowmelt, higher evapotranspiration in early summer, and earlier spring peak flows, leading to reductions of 75 to 90 percent in April-September runoff volumes; studies of direct climate change effects on hydrology for Montana and California produced similar results (Running and Nemani, 1991; Lieth and Whitfield, 1998; Miller et al., 2003; Dettinger et al., 2004). These changes are expected to exacerbate conflicts over limited dry season river flows among energy production, irrigation, instream flow, and recreational uses (Hamlet and Lettenmeier, 1999). Simulated hydrologic effects of climate change show an increase in competition for reservoir storage between hydropower and instream flow targets developed in response to the Endangered Species Act listing of Columbia River salmonids (Payne et al., 2004). In the Colorado River basin, simulated future climate scenarios show a reduction in water storage, reduced hydropower production, and an increase in the number of years in which reservoir releases do not meet demand (Christensen et al., 2004).

These predictions of direct effects of climate change on water yield from forested basins do not take into account many potential indirect and interacting effects of forest responses to climate change (Clark, 1990; Stocks et al., 1998; Dale et al., 2001, Walther et al., 2002). For example, Westerling et al. (2006) assert that climate warming is responsible for the increased frequency of wildfires and longer fire seasons in the western United States. Widespread outbreaks of pine beetle in British Columbia and the Rocky Mountains also are attributed to climate warming (Taylor and Carroll, 2004; Hicke et al., 2006). By altering forest disturbance, climate change may indirectly affect water yield and water quality.

In addition to effects on forest disturbances, climate change is expected to alter forest productivity and species composition (Aber et al., 1995, Houghton, 1995; USGCRP, 2000). Forest productivity will change (Melillo et al., 1993). Forest species composition is changing (Brown et al., 1997; Davis and Shaw, 2001; Pearson and Dawson, 2003; Parmesan and Yohe, 2003; Thomas et al., 2004). By altering the productivity and species composition of forests, climate change may indirectly modify water quantity and quality.

Research Need: Studies of indirect and interacting effects of climate change on water quantity and quality through effects on forest disturbance, structure, and species composition.

SOCIAL RESEARCH NEEDS

The social context of forest and water interactions has changed since the mid-twentieth century and continues to shift today. Current issues involving forests and water encompass practices that extend beyond the traditional scope of timber production and now include multiple public and private groups, past and future land use trends, and non-market resource valuation and trading schemes. Changing forest landscapes also include rapid changes in the public policy setting. Today, with growing populations in or adjacent to forestlands and more stringent polices regulating forest management, there are many more groups of people influencing where and how forests should be grown or preserved, and this influences forest water resources.

Urban and Exurban Development

Future expansion of homes, commerce, and industry replacing forests in urban, suburban, and exurban areas is likely to produce hydrologic effects. In the last few decades of the twentieth century, "exurban sprawl" (sprawl development in the farthest fringes of metropolitan areas) changed demographic patterns (Alig, 2006). Compared to urbanization, sprawl is more diffuse, is more widespread, and affects more area per unit of population. Through exurban sprawl, human populations spread into rural areas, extend the amount of impervious area, and fragment remaining blocks of forest. Across the United States, the proportion of area classified as rural area has declined in most counties, housing has fragmented many large forest patches, and housing density has increased over large exurban areas (Theobald, 2005; Goetz et al 2004; Auch et al., 2004).

Continued exurban sprawl is expected to reduce forest cover throughout the United States, with potential consequences for water yield and quality for municipalities. In the western United States, census data from 1960 to 2000 show development spreading, especially along the coast (Travis et al., 2005). By 2040 a large swath of development is projected along the west coast, and the footprints of larger interior cities such as Phoenix, Denver, and Salt Lake City are projected to increase. Rural valleys across the mountain West—for example, in western Colorado—exhibit marked exurbanization by 2040 (Travis et al., 2005; Theobald, 2005). The trend is evident in the eastern United States, too. Population increases and exurban sprawl are occurring in the south and southeastern United States (Theobald, 2005), New England (Foster et al., 2004), and the mid-Atlantic United States (Goetz et al., 2004). Direct consequences of these patterns are an increase in impervious area (faster runoff), and with more wildland-urban interfaces, houses come into greater contact with wildland processes, including windthrow of trees, landslides, and fire (Radeloff et al., 2005). Much is known about the localized and larger-scale effects of urbanization on hydrology, but it is not clear whether the hydrologic effects of exurban sprawl can be predicted by these past studies because of differences between sprawl and urbanization.

The new patterns of sprawl and development in forested landscapes create an opportunity for new interdisciplinary studies involving hydrologists, ecologists, economists, and social scientists to improve and communicate understanding of the value of forests in their role of producing water. In Arizona, economists worked with hydrologists to translate biophysical responses to treatments into production functions that capture economic impacts of forest change on hydrology (Baker and Ffolliott, 1998). Interdisciplinary research by hydrologists, ecologists, managers, and economists can help to translate research and monitoring results into economic terms that have meaning to decision makers and policy analysts.

Research Needs:

- **Landscape-scale analyses that assess the effects of exurban sprawl on water quantity and quality.**

- **Collaborative research among hydrologists, ecologists, and economists and social scientists to improve and communicate understanding of the value of sustaining water resources from forests.**

Changing Forest Practices on Public Lands

Environmental laws have led to reduced timber harvest on public forestlands, and wildfires appear to be increasing in severity and extent. Yet, simultaneously, more people have moved into the urban-wildland interface in or near forests. These factors have turned attention to protecting people and their property from forest disturbances, such as fire, landslides, and wind storms. Some contemporary management practices, therefore, cater to these new social conditions and involve new forms of timber harvest whose effects on hydrology are not well understood.

Federal forests are managed for a wide range of objectives (see Chapter 2). Today, many of the federal forest lands are managed to conserve terrestrial and aquatic species and protect water, practices that constrain the amount and extent of timber harvest (USDA and USDI, 1994; Christensen et al., 1996; Thomas, 1996). On public lands, legal requirements for species protection, forest preservation, and fire in effect limit forest management options to a very narrow scope. One noted example of contemporary management practices is the Northwest Forest Plan (1996), which restricts cutting of mature and old-growth forests and mandates wider riparian buffers to protect the habitat of an endangered species, the spotted owl. After fire, salvage logging may occur on public forest lands (Donato et al., 2006; Thompson and Spies, 2007). Ongoing road decommissioning has occurred in key watersheds of the Aquatic Conservation Strategy of the Northwest Forest Plan (Box 2-1). Other contemporary forest management practices derived from the Healthy Forests Restoration Act (2003), including thinning for fuel reduction and forest restoration. In western forests where conflicts arise over the allocation of limited summer low flows between endangered species, agriculture, and other uses, managers need to understand the effects of contemporary practices, such as various levels of forest thinning, buffers, and salvage logging, on streamflow and aquatic ecosystems.

Research Need: Studies of effects on water quantity and quality of contemporary forest management on public lands, including thinning for fuel reduction or forest restoration, salvage logging, road decommissioning, and redesigned riparian buffers.

CUMULATIVE WATERSHED EFFECTS

Cumulative watershed effects (CWEs) include any changes that involve watershed processes and are influenced by multiple land use activities. Assessments of CWEs use interdisciplinary approaches at the large watershed scale and attempt to include longer temporal scales (Reid, 1993; MacDonald, 2000). Assessing CWEs requires an understanding of physical, chemical, and biological processes that route water, sediment, nutrients, and other pollutants from uplands to downstream areas (Sidle, 2000).

Research on CWEs attempts to establish cause-effect relationships among forests and water over large spatial scales; however, CWEs are elusive to quantify and are especially difficult to convey in terms applicable to policy and management. Many of the methods for evaluating cumulative effects have encountered technical, legal, or political problems because they have not explicitly addressed the complexity of biophysical interactions spread over large areas and long time frames (Grant and Swanson, 1991).

Spatially explicit studies of hydrologic responses to forest management display how land uses and land cover across large watersheds interact to influence the quantity, timing, and quality of streamflow. In many parts of the United States, novel analyses represent watershed processes in large watersheds and landscapes. Three case studies are presented: Oregon (Box 4-1), the Chesapeake Bay (Box 4-2), and New England (Box 4-3). Each of these cases reflects: (1) satellite image interpretation of land cover in heterogeneous watersheds; (2) public participation in envisioning future scenarios in the watershed; and (3) spatially explicit alternative future land cover patterns and engineered features using geographic information systems and spatial models (Baker and Landers, 2003; King et al., 2005).

In the first example (Box 4-1 and Figure 4-4), the Environmental Protection Agency (EPA) funded work to reconstruct historical forest cover and project future forest cover under alternative scenarios for future development in the forested Willamette River drainage basin, a 30,000 km^2 area in Oregon that includes the largest cities in the state. In the Chesapeake Bay drainage basin, regional coordination and large watershed-scale modeling are used to help the five states in the basin to meet nutrient reduction goals; forests are a critical part of each state's strategy, particularly riparian forest buffers and conserving existing forests, and state goals (Box 4-2 and Figure 4-5). Finally, the New England case (Box 4-3 and Figure 4-6) shows forest management on public land for the purpose of protecting municipal drinking water in a 75 km^2 watershed in Connecticut and Massachusetts by limiting harvest in forests close to the reservoir on steep slopes or unstable soils.

While all of these examples display the power and effectiveness of geospatial analysis and ways to combine, analyze, and communicate complex data, none of them explicitly focuses on CWEs or water quantity and quality. However, all three of these studies represent the spatial patterns of forests in large watersheds, implicitly draw on principles of forest hydrology, and aggregate scientific principles relevant to policy making at the large-watershed scale. These studies illustrate the potential for forest hydrologists to use geospatial and geostatistical tools to analyze and display hydrologic processes in large, heterogeneous watersheds, and they are excellent examples of how forest hydrology and spatially explicit CWE research could be done. With these types of tools—including the rapid advancements in and increased availability of spatial data, greater power of geographic information systems, and gains in scientific knowledge—forest hydrology could develop a reasonable path forward to assessing CWEs and fulfilling other needs of forest management in the twenty-first century.

Research Need: Spatially explicit assessments of CWEs in large watersheds that connect and communicate watershed processes, changing land cover, management issues, and public participation.

SUMMARY

Expanding from general principles (Table 3-1) of hydrologic responses to forest management and disturbance (Table 3-2) to predicting responses in changing forest landscapes will involve meeting research needs that extend the spatial and temporal scales and social considerations in forest hydrology. These extensions to the existing rich body of forest hydrology science can improve the application of forest hydrology to address forest management issues in the twenty-first century. Most forest hydrology research has been conducted at small spatial scales, over short time scales, and in watersheds with homogeneous land cover. Today, forest and water managers need predictions of direct, indirect, and interacting hydrologic responses to changing forest landscapes and guidance in applying these predictions at the scales of large watersheds, landscapes, and regions, over multiple decades. Spatial, temporal, and socioeconomic factors and climate are important sources of change in forested landscapes, and each of these has research needs and challenges associated with improving forest hydrology applications to manage forests for forest and water resources. This chapter describes the challenges and the research needed to address them. Cumulative watershed effects are discussed and examples are given of how the science of forest hydrology can use existing technologies to quantify and communicate hydrologic effects over large spatial scales and in basins with heterogeneous land cover.

BOX 4-1
Integrating Stakeholder Perspectives to Manage Future Forest Landscapes in the Willamette Valley, Oregon

The Willamette Basin Alternative Futures Analysis is a novel environmental assessment approach (*http://oregonstate.edu/dept/pnw-erc/index.htm*) that facilitates consensus building and helps communities make decisions about land and water use (EPA, 2002). It combines technological capacities for reconstructing past landscapes and modeling future landscapes with public consultation to provide a long-term, large-area perspective on the combined effects of multiple policies and regulations affecting the quality of the environment and natural resources within a geographic area. In this process, community members articulate and understand their different viewpoints, priorities, and goals.

The Willamette River drains an area of nearly 30,000 km^2 between the Cascade and Coast Range Mountains in western Oregon (see Figure 4-4). Forests occupy two-thirds of the basin, mostly in upland areas, while much of the lowland valley area has been converted to agricultural use (43 percent) and urban and rural development (11percent). Oregon's three largest cities, Portland, Salem, and Eugene-Springfield, are located in the valley, along the Willamette River. About 2 million people lived in the basin in 1990. By 2050, the basin population is expected to nearly double, placing demands on land and water resources and creating challenges for land and water use planning. In the mid-1990s, recognizing the need for an integrated strategy for development and conservation, Oregon Governor John Kitzhaber initiated basinwide planning efforts and created the Willamette Valley Livability Forum (*http://www.wvlf.org*).

Working with stakeholders, researchers outlined three alternative futures and projected them through the year 2050. Plan Trend 2050 represents the expected future landscape based on current policies and recent trends; Development 2050 reflects a loosening of current policies to allow freer rein to market forces; and Conservation 2050 places greater emphasis on ecosystem protection and restoration. All three futures assume the same population increase. The historical, present-day, and future landscapes are represented as maps (see Figure 4-4) with associated assumptions about management practices and water use and computer-simulated "flyovers" of the future conditions. Researchers used models to compare expected effects of the alternative futures on terrestrial wildlife, water availability, ecological conditions of streams, and the condition of the Willamette River (Baker and Landers, 2003; Dole and Niemi, 2003; Hulse et al., 2003; Van Sickle et al., 2003).

Major changes to the Willamette River Basin since EuroAmerican settlement in 1850 include: (1) loss of 80 percent of riparian forests; (2) conversion of about two-thirds of the old-growth forest in the uplands to younger forest; (3) drying of an estimated 130 km of second- to fourth-order streams in a moderately dry summer due to consumptive water use for irrigation, municipal, industrial, and other out-of-stream water uses; and (4) 15 to 90 percent declines in wildlife habitat and abundance, and stream and river biota.

Projected conversion of agricultural lands to rural and urban development between 2000 and 2050 produced smaller effects on ecosystems than conversions of riparian or upland forest to either agriculture or urban land uses. All three futures involved substantial increases in water use and declines in water availability, resulting in habitat loss and summer drying of streams. Changes were greatest in the Plan Trend and least in the Conservation future, but even the water conservation measures incorporated into Conservation 2050 were not sufficient to reverse recent trends of increasing water withdrawals for human use. Researchers concluded that major changes in Oregon's water rights laws would likely be needed to substantially reduce water withdrawals, but stakeholders did not consider such changes to be plausible.

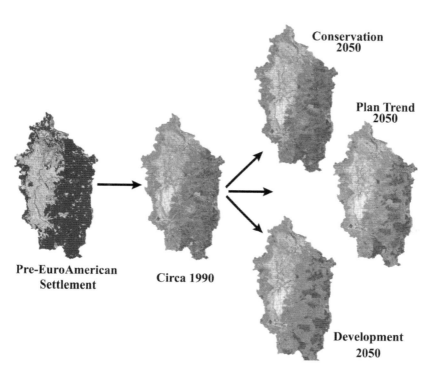

FIGURE 4-4 Trajectories of landscape change in the Willamette River Basin, from pre-EuroAmerican settlement, to ca. 1990, to three alternative futures for 2050. Source: EPA (2002).SOURCE: Available online at *http://oregonstate.edu/dept/pnw-erc/index.htm*.

BOX 4-2
How to Keep Forests on the Landscape: The Chesapeake Bay Case Study

Once a primarily forested landscape, the Chesapeake drainage has undergone several centuries of settlement and conversion to agricultural and urban uses. Forests still cover 58 percent of the basin, although the area now supports more than 16.6 million people (Figure 4-5). Despite their extent, forests are estimated to contribute only 14 percent of the nitrogen and 2 percent of the phosphorus in the basin. Hence, the issues for forest hydrology become how to keep forests on the landscape and where they are most critical for maintaining basic environmental functions. Forest retention, especially in hydrologically active areas such as riparian zones, is an important practice for mitigating the effects of other land uses.

The Chesapeake Bay has been the focus of multistate coordination and widespread efforts to reduce nutrients since the 1980s, while population grew by 23 percent. With larger houses and more roads to serve lower-density development, impervious surfaces expanded at even higher rates, creating effects on water quantity and quality even where some stormwater measures were in place to mitigate the increases in peak flows. Greater total flow from impervious areas enlarged stream channels, mobilizing sediment stored from previous decades of agricultural erosion or abandoned mill dams. With less water infiltrated and slowly released from subsurface soils, summer baseflows in developed areas declined. New nutrient loads were added through more intensive management for turf and landscaping and septic or sewer systems. Although some measures of water quality have improved measurably in portions of the watershed since 1983, the waters remain stubbornly below standards set for clarity, dissolved oxygen, and other criteria. The stakes are going up as the deadline to meet and maintain water quality standards for the Chesapeake Bay mainstem nears, with Total Maximum Daily Load limits slated for 2010 in a watershed that ranges over six states. The low rates of nutrient export from forests will be needed to meet water quality standards, as well as support wildlife and aquatic habitats dependent on the forests.

The effects of development and agriculture usually have strong signals in water quality and habitat measures (e.g., King et al., 2005). Hydrologic changes associated with urbanization are even larger. In contrast, the changes from forest management can be difficult to distinguish from annual variation, at least where practices such as stream buffers and appropriate road design are in place (McCoy et al., 2000). Where the effects of forest management on water quality and quantity are measurable, they are generally at least an order of magnitude less than those from other land uses.

Forests are used in a variety of ways to mitigate the adverse effects of land use change in the Chesapeake Bay basin. The Chesapeake Bay Program adopted an ambitious riparian forest restoration goal of 10,000 miles in 2003. A 1996 goal of 2,010 miles by 2010 was met years ahead of time as a result of the new Conservation Reserve Enhancement Program in Maryland, Virginia, and Pennsylvania. Efforts are under way to increase forest conservation significantly. Forests retained on development sites are allotted credits towards meeting stormwater requirements in Maryland. "Critical area" laws in Virginia and Maryland limit density and forest clearing, especially in buffer areas, and require replacement of cleared forests. Urban canopy cover goals are being set to benefit from air and water quality improvements of trees in developed areas (Cappiella et al., 2005). As forest area continues to decline, forests increasingly are being retained and restored to mitigate the effects of other land uses on water quantity and quality.

Forests are seen by some to be the vacant part of the landscape, waiting for a higher and better use. However, forests are essential to maintaining basic environmental functions, and explicit methods to maintain forest on the landscape will be needed to meet basic water quality goals.

FIGURE 4-5 Classification of land cover west of the Chesapeake Bay. White and red areas are urban and impervious zones, yellow is agriculture, and green is forest. SOURCE: Available online at *http://www.geog.umd.edu/ resac/lc2.html*. Reprinted, with permission, from Stephen D. Prince (May 19, 2008). Copyright by the Department of Geography, University of Maryland

BOX 4-3
Forest Management for Watershed Protection: New England

The Barkhamsted Reservoir watershed drains an area of about 75 km^2 of Connecticut and Massachusetts, including private lands and public forestlands owned and managed by the Metropolitan District Commission (MDC) and Connecticut and Massachusetts state forests. Forests in the watershed are managed for timber production and other uses, but these uses have the potential to adversely affect water quality and quantity in the reservoir. To protect the municipal water resources, a decision support system (the Watershed Forest Management Information System), has been developed to

1. map the forest and land cover and engineered features such as roads in the watershed using geographic information systems;

2. designate

- forests and wetlands for conservation based on their perceived role in supplying clean water (the Conservation Priority Index [CPI]);
- agricultural lands and parks for restoration; and
- residential, commercial, and industrial lands for nonpoint source pollution;

3. identify

- roads as sediment sources, according to their proximity to water bodies; and
- culverts for failure given estimated peak discharges; and

4. spatially allocate and schedule forest harvest and silvicultural operations by position within the watershed (Barten and Ernst, 2004).

Using this system of indices, each parcel is given a score that represents its conservation value within the watershed (Figure 4-6). Because the information is shared among all landowners in the basin, scores can be used to build cooperation among state or federal agencies and nongovernmental organizations to focus conservation efforts. Watershed sensitivity classes constructed from the CPI (Figure 4-7) can be used to delineate and coordinate forest harvests. Forests around water bodies and wetlands are in the highest-sensitivity class to facilitate stream habitat, forest diversity and forest regeneration. This approach illustrates the potential for interagency cooperation, coordination, and information sharing to guide watershed management plans in large, multi-ownership watersheds.

SOURCES: Barten and Ernst (2004); Zhang (2006).

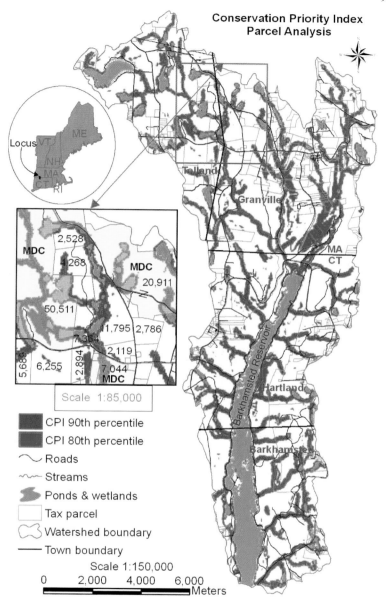

FIGURE 4-6 Conservation priorities assigned to forests on public and private lands within the Barkhamsted Reservoir. SOURCE: Gregory, P.E., Y. Zhang, and P.K. Barten. Watershed Forest Management Information System (WFMIS) User's Guide Version 1.0. USDA. Available online at http://www.wetpartnership.org/WFMIS%20User%27s%20Guide.pdf.

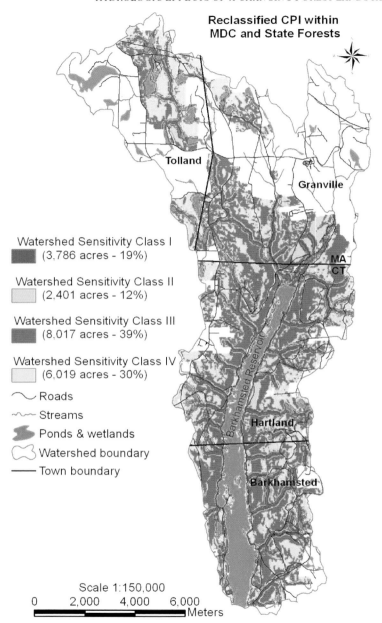

FIGURE 4-7 Public lands classified by watershed sensitivity (class I is highest sensitivity, most restricted use; Class IV is lowest sensitivity, least restricted use) in Barkhamsted Reservoir Watershed. SOURCE: Gregory, P.E., Y. Zhang, and P.K. Barten. Watershed Forest Management Information System (WFMIS) User's Guide Version 1.0. USDA. Available online at http://www.wetpartnership.org/WFMIS%20User%27s%20Guide.pdf.

5
Recommendations for Forests and Water in the Twenty-First Century

The preceding chapters have outlined the working understanding of (1) forests, forest management, and emerging issues facing forests (Chapter 2); (2) the state and limits of the body of forest hydrology science (Chapter 3); and (3) research needs to meet management challenges in changing forest landscapes (Chapter 4). Common themes thus far in the report include rapid changes in forests and water systems, science, and management; fragmentation in technology and information transfer across the scientific, management, and citizen communities; and the need to apply scientific principles to larger spatial and longer temporal scales. This chapter builds on previous chapters to recommend actions for scientists, managers, and citizens to better understand connections between forests and clean, plentiful water and to use that understanding to promote sustained water resources from forests. These recommendations are structured to begin to bridge some of the fragmented elements across scientific research, management policies, and community activities.

RECOMMENDATIONS FOR FOREST HYDROLOGY SCIENTISTS

Managers and citizens will make policy and land use decisions about forests and water based, in part, on scientific knowledge, which elevates the role of scientists in those decisions. Forest hydrology plot, process, modeling, and watershed studies provide the foundation for water and forest resources for this and subsequent generations. Recommendations for scientists to meet water and forest needs fall into three categories: maintaining and enhancing watershed studies, incorporating emerging technologies in research, and developing models for addressing management needs in an uncertain future.

Maintaining and Enhancing Small Watershed Studies

The combination of small watershed and process studies has built a solid foundation of forest hydrology science (Stednick et al., 2004). Over the last half of the twentieth century, small watershed studies collected hydrologic data from forests in many geographic regions, and some of these data span multiple decades. For the oldest records, efforts may be needed to transcribe analogue or hand-written records into digital formats. No matter the form, data records from all small watersheds are of great value for scientists and managers because they provide a foundation for long-term monitoring as well as a collective database that can be used for meta-analyses of the effects of forest change on runoff (e.g., Jones, 2000).

Long-Term Monitoring Data

Monitoring involves repeated data measurements that are used to detect changes or trends over time. Two types of monitoring are useful to forest hydrology: (1) continuous or repeated measurements of streamflow, stream temperature, and stream chemistry; and (2) records obtained over an area, such as by aerial photography or satellite imagery. As these data accumulate over time, they greatly increase in value. Long-term monitoring permits observation and contextualization of extreme hydrologic events; detection of trends or cycles in climate or vegetation; and assessment of long-term responses to experimental treatments or disturbance (e.g., Stednick et al., 2004; Jones, 2005). Monitoring records from nested watersheds (i.e., from headwaters to large watersheds) can reveal cumulative watershed effects. Finally, monitoring data can be used to validate models, including those that predict responses to extreme events.

Pieces of the infrastructure for long-term monitoring are already in place. Small watersheds exist on public land, including properties maintained by the U.S. Forest Service, the Agricultural Research Service (ARS), the U.S. Geological Survey (USGS), the Department of Energy (DOE), and the National Park Service. Additional hydrologic data are available from the Environmental Protection Agency. In addition, the U.S. Geological Survey has a national streamgaging system that collects hydrologic flow statistics on streams across the country, and many of these gaging stations have multiple decades of streamflow information. Of serious concern is that many long-term monitoring efforts have been, or are at the risk of being, suspended due to funding and other constraints.

Maintaining existing small watershed studies and reestablishing data collection at abandoned sites could help address key questions about the long-term hydrologic effects of forest change and conversions. Resurrected monitoring and data collection activities could provide information on measurable hydrologic effects at abandoned experimental watersheds and monitoring stations that have experienced fires or insect infestations since data collection ceased. For example, reestablishing measurements at the historic, experimental Wagon Wheel Gap watersheds in Colorado could allow comparisons of earlier data on forest cover change with data representing the ranching and second-home development that now exists in the watershed. The Watts Branch watershed in Maryland, where Leopold and colleagues (Leopold et al., 1964) conducted their fundamental studies of channel-forming flows, has subsequently undergone extensive land use change and downcutting of the channel. Recent historical reconstructions (Walter and Merritts, 2008) indicate that downcutting of Watts Branch and many other eastern streams was due to early to mid-nineteenth century breaching of small mill dams that had impounded sediment eroded from forest conversion to agriculture during the 1700s and 1800s. Reestablished streamflow monitoring at sites such as Wagon Wheel Gap and Watts Branch could improve understanding of streamflow responses to complex changes in land use and natural disturbance.

In places where data collection has been reestablished on small watersheds, it has produced valuable insights. For example, small instrumented watersheds in the ponderosa pine forests of central Arizona were reinstrumented after the Rodeo-Chediski fire of 2002, and these new data allowed comparison of post-fire effects to pre-fire conditions of the 1970s (Ffolliott and Neary, 2003). Also, small instrumented watersheds in Coyote Creek, Oregon, were reinstrumented 35 years after forest harvest treatments, and these new data allowed detection of long-term forest regeneration and fire suppression effects on streamflows (Perry, 2007).

Small Watershed Data as a Meta-Experiment

One of the great values of forest hydrology science in small watersheds during the twentieth century lies in the power of the collective dataset, which spans broader spatial scales, research goals, and geographic regions than any individual site. New analyses of the various data in this collective dataset could treat the entire collection of small watershed data as a "meta-experiment." This meta-experiment would require a new approach to data analysis and could be structured to address some of the research and management questions that span large spatial scales or long time periods. Using the data in this way could extend the familiar individual, small watershed studies to better understand connections between changing forest processes and watershed responses.

A meta-experiment of forest hydrology from small watersheds could yield clearer understanding of long-term changes in forested "control" watersheds in response to fire suppression, climate change, and land use across different sized watersheds (Jones, 2005). Multiple agencies (USFS, ARS, Environmental Protection Agency [EPA], DOE, and USGS) have historical records from early experiments that have not been digitized, as well as long-term records that are available online (e.g., the U.S. Forest Service's Clim-DB/HydroDB project [*http://www.fsl.orst.edu/climhy/*] or the USGS National Water Information System web site [*http://water.usgs.gov/data.html*}). This next generation of watershed analyses could be undertaken as a multiple agency effort to (1) reestablish monitoring at key sites that have valuable early records, but have been abandoned; (2) digitize historical streamflow and other monitoring records; (3) gather data in centralized locations and make them available online; and (4) develop automated methods for comparison of long-term records, using computer-based techniques. These data could then be analyzed as a collective set to detect changes across many watersheds and to improve understanding of connections among hydrology, forest systems, land use, management, climate variations, and time.

Recommendations for Small Watershed Studies

- Scientists should continue small watershed experiments and studies and reestablish monitoring at key sites where data collection and monitoring activities have ceased;
- Scientists should centralize historical records from watershed studies in

digital, well-documented, publicly accessible databases;
- Scientists should use the entire collection of small watershed studies as a meta-experiment to increase understanding of forest hydrologic processes; this effort would involve the following elements:
  — Gathering all data in centralized locations that are available online;
  — Developing automated methods for comparison of long-term records, using current computer-based techniques;
  — Examining long-term "control" watershed variability and response to timber harvest, fire suppression, climate change, and disturbances; and
  — Investigating effects of hydrologic changes on aquatic ecosystems.

Emerging Technologies for Quantitative Analysis

In addition to continuing and enhancing paired watershed studies, some new and emerging technologies can help advance forest hydrology. Emerging technologies relevant to forest hydrology include (1) satellite imagery and remote sensing, (2) distributed sensor networks, and (3) geographic information systems (GIS) and associated geostatistical and visualization techniques. These emerging technologies make possible data collection over large spatial areas that could be used in combination with plot and process data to better understand how measurements in the experimental sites compare to unmeasured areas in the same watersheds or at the landscape scale (see Box 5-1).

Remote Sensing Technologies

Airborne and satellite remote sensing techniques can save time and costs associated with monitoring hydrologic processes over large watersheds and regions, including estimation storage of water in the atmosphere, snow, vegetation, and soil, as well as evapotranspiration (ET). New satellites such as the Terra and Aqua, which both carry the Moderate Resolution Imaging Spectroradiometer (MODIS) sensor, are viewing the entire Earth's surface every one to two days, acquiring data in 36 spectral bands (groups of wavelengths). Currently available MODIS image products provide daily estimates of ET at 1 km resolution. Other satellites operating in both the optical and the microwave parts of the spectrum are useful for mapping areas of inundation and saturation (Toyra et al., 2001; Sass and Creed, 2007; Clark and Creed, submitted). MODIS, with scaling techniques to reconcile differences in resolution, is being used to provide distributed water balance information at fine spatial and temporal scales (Singh et al., 2004).

LiDAR (Light Detection and Ranging) is another emerging technology relevant to forest hydrology. LiDAR imagery has become increasingly available in recent years. Raw LiDAR data can be processed in different ways. In their most common form, LiDAR data are processed into topographic data, and in this form, LiDAR has

> **BOX 5-1**
> **Applications of Diffuse Reflectance Spectroscopy and Stable Isotopes to Monitor Landscape Features and Environmental Services**
>
> Scientists from the World Agroforestry Center, headquartered in Nairobi, Kenya, and colleagues have developed and applied diffuse reflectance spectroscopy and isotope methodologies to make rapid assessment of the impacts of deforestation and other land use changes on ecosystem properties, soil organic carbon, and soil quality (fertility). Areas of sediment deposition have been linked to source areas of sediment from upland soil erosion. The technology presently requires that reference soil spectral libraries be developed from soil samples obtained from the watersheds of interest. "Reflectance fingerprints" are obtained that can quantify and simultaneously predict multiple soil and plant properties. These technologies have promise for applying remote sensing to assess the spatial condition of soils and vegetation that control water flow and water quality.
>
> SOURCES: Shepherd and Walsh (2002, 2004); Vagen et al. (2005).

revolutionized the mapping of topography at fine spatial scales. In a less commonly used form, these data can be processed into forest canopy structure (Lefsky et al., 2002). In this form, the data have an unrealized potential to provide insights as to how canopy structure affects forest hydrologic processes. Through MODIS and LiDAR, essential data that were previously unavailable or difficult to obtain can now be used to model and more accurately predict water storage and flows through large watersheds and regions.

Distributed Sensor Networks

Multisensor networks connected through wireless technology are under rapid development. These networks are based on nanotechnology and can inexpensively measure key variables such as soil moisture and temperature, at high spatial and temporal resolutions (Szewczyk et al., 2004). At this point in their development, novel sensor networks are limited by basic engineering constraints, such as power sources and technical challenges of processing large amounts of data. Current efforts to establish and test sensor networks are focusing on the plot or small watershed scale, but they are not yet being implemented at larger spatial scales. Still, sensor networks hold promise and possibilities to greatly improve understanding of hydro-ecologic processes at fine scales.

Geographic Information Systems and Geovisualization

Over the past few decades, developments in geographic information systems, including global positioning systems (GPS), digital elevation models (DEMs), and computer capabilities have greatly advanced the ability to collect and analyze very large spatial and temporal datasets (Guertin et al., 2000). GIS has been shown to be a powerful tool, especially when combined with spatial modeling, to predict water and sediment transport in small and large watersheds. For example, the NetMap system (Benda et al., 2007) predicts erosion potential, sediment supply, road density, forest age, fire risk, hillslope failures, and stream habitat indices. Models with watershed terrain analysis features can facilitate planning and management such as targeting "ecological hotspots" for stream restoration. The results can be useful for comparing alternative management scenarios and assessing cumulative watershed effects (CWEs). These technologies are progressing rapidly, allowing new types of predictions at finer resolution and over larger areas than were thought practical even a decade ago.

Recommendations for Emerging Technologies

- Scientists should refine GIS, remote sensing, and sensor networks to increase understanding and prediction accuracy of hydrologic responses at large watershed scales.
- Forest hydrologists should be trained to understand and use new GIS, remote sensing, and sensor network tools for forest hydrology applications or should develop effective collaborations with specialists.

Hydrologic Models

Models are important tools for scientists to predict, simulate, and compare the effects of different controlling factors from theory, experiments, and observations across various spatial and temporal scales. In forest hydrology, numerous hydrologic models have been developed for many different objectives (Singh, 1995; Singh and Frevert, 2006), such as determining the size of culverts for roads or predicting the hydrologic impacts of land use change over different spatial scales and time periods. These models vary in how they represent hydrologic processes, vegetation, soils, groundwater, and runoff; they also vary in the spatial and temporal scales at which they simulate hydrologic processes.

Many hydrologic models have limited capacity to simulate the hydrologic processes and response of natural and altered forested watersheds. One important limitation is due to an implicit assumption that overland flow is the dominant cause of runoff in forested watersheds (Dunne and Leopold, 1978; Hawkins, 1993; Hawkins and Khojeini, 2000; Eisenbies et al., 2007) and that models of sediment production are predicated on these overland flow models (Renard et al., 1997; Williams, 1995; Neitsch et al., 2002; Boomer et al., 2008). Another limitation is that most hydrologic models are developed and validated at the spatial scale of small watersheds, and it is difficult to connect forest hydrologic models to models at the large-watershed scale, or to regional or global climate models.

Models have great potential to represent and communicate hydrologic effects and CWEs, but scientists differ on how to approach this challenge. Many modelers agree that physically based models are more illuminating than models based on empirical relationships (Sidle, 2006). However, advances are needed to develop physically based models at larger spatial and longer temporal scales than the small watershed scale at which models are typically developed and tested. Ideally, physically based models would be based on data and parameters that forest and water managers monitor. Large-scale monitoring using new technologies and long-term monitoring of watersheds can provide some basis for developing scaling rules. The research needs for advancing forest hydrology science include understanding long-term and landscape-scale hydrologic effects of fire and fire suppression, climate change, and cumulative watershed effects (see Chapter 4). Spatially explicit assessments and physically based models designed to simulate, predict, or represent these phenomena form the basic needs of forest hydrology-related models for today and the foreseeable future.

Recommendations for the Next Generation of Hydrologic Models for Forest Hydrology Applications

- Forest hydrologists should extend the capability of models to incorporate the kinds of changes happening in forests, such as fire, cumulative watershed effects, and climate change;
- Forest hydrologists should advance models to simulate hydrologic processes across large watersheds; and
- Forest hydrologists should use emerging technologies and long-term datasets to build and test the next generation of forest hydrology models.

RECOMMENDATIONS FOR MANAGERS

Evolving Best Management Practices

Best management practices (BMPs) are widely used to prevent or reduce the negative hydrologic effects of forests and land use activities (see Box 5-2). Forestry BMPs are forest management practices intended to mitigate the negative consequences of timber harvest, road construction and maintenance, reforestation, or other forest management practices (Binkley and Brown, 1993; Seyedbagheri, 1998; Aust and Blinn, 2004). BMPs are employed in forests of many different ownerships across the United States.

Although "best" connotes an ideal condition or superior approach, in fact, BMPs are most often negotiated compromises between parties with economic interests in management activities and those with interests in environmental protection. The balance between these two continually evolving sides is a "best" compromise. The environmental side of this negotiation is required by key pieces of legislation, such as the Clean Water Act (CWA) of 1972. A number of studies have assessed compliance with and effectiveness of BMPs with respect to 1970s goals for environmental protection, such as reducing nonpoint source pollution (Binkley and

> **Box 5-2**
> **BEST MANAGEMENT PRACTICES**
>
> Best management practices (BMPs) are effective, practical, structural or nonstructural methods that prevent or reduce the movement of sediment, nutrients, pesticides, and other pollutants from the land to surface or groundwater, from nonpoint sources such as silvicultural activities (Brown et al., 1993; Brooks et al., 2003; Chang, 2003).
>
> The Federal Water Pollution Control Act Amendments of 1972, Public Law 92-500 (and as amended by Section 319, 1986) require the management of nonpoint sources of water pollution from sources including forest-related activities. BMPs have been developed to guide forest landowners, other land managers and timber harvesters toward voluntary compliance with this act. A central objective of this law is to maintain water quality to provide "fishable" and "swimmable" waters. The Environmental Protection Agency (EPA) recognizes the use of BMPs as the primary method of reducing nonpoint source pollution.
>
> Nonpoint sources of pollution are diffuse and may include fertilizers, herbicides, and insecticides from agricultural lands and residential areas; oil, grease, and toxic chemicals from urban runoff and energy production; sediment from construction sites, crop, and forestlands, and eroding stream banks; salt from irrigation practices and acid drainage from abandoned mines; and bacteria and nutrients from livestock, pet wastes, and septic systems. The amounts of pollutants from single locations often are small and insignificant, but when combined over the landscape, they can create water quality problems. The adoption and use of BMPs help achieve the following water quality goals:
>
> 1. To maintain the integrity of stream courses;
> 2. To reduce the volume of surface runoff originating from an area of forest management disturbance and running directly into surface water;
> 3. To minimize the movement of pollutants (e.g., pesticides, nutrients, petroleum products) and sediment to surface and ground water; and
> 4. To stabilize exposed mineral soil areas through natural or artificial revegetation.
>
> Although it is unrealistic to expect that all nonpoint source pollution can be eliminated, BMPs can be used to minimize the impact of forestry practices on water quality. A thorough understanding of BMPs and flexibility in their application are of vital importance in selecting BMPs. More than one BMP may be effective for a given situation. BMPs usually are designed to be practical and economical while maintaining both water quality and the productivity of forest land.
>
> SOURCES: Hawaii Watershed and Management Program (*http://www.state.hi.us/dlnr/ dofaw/wmp/bmps.htm*) and Environmental Protection Agency (*http://www.epa.gov/watertrain/ forestry/ forestry3.htm*).

Brown, 1993; Aust and Blinn, 2004). In these studies, compliance with BMPs varied from 30 to more than 90 percent, with lower levels of compliance associated with road and trail decommissioning and higher levels of compliance with riparian buffer strips (Briggs et al., 1998; Schueler and Briggs, 2000). Studies in the eastern United States have shown that BMPs significantly lowered nonpoint source pollution (sediment, temperature, some nutrients) from clear-cuts and roads (Lynch et al., 1985; Lynch and Corbett, 1990; Kochenderfer et al., 1997; Arthur et al., 1998; Wynn et al., 2000; Vowell, 2001; Aust and Blinn, 2004). Studies indicate that

BMP effectiveness is site-dependent (Blinn and Kilgore, 2001; Broadmeadow and Nisbet, 2004; Lee et al., 2004), although some general trends emerge. Assessments conducted in the 1980s and 1990s give high marks to BMPs, such as riparian buffer strips for effectiveness in reducing local sediment contributions to streams and other forms of nonpoint source pollution (Ice 2004). However, very little research has investigated whether the current suite of BMPs will be effective in reducing cumulative watershed effects, maintaining viable fish populations, or preserving the integrity of forest and stream ecosystems (Swanson and Franklin, 1992; Bisson et al., 1992, Ice 2004).

Forest and watershed managers have an important role to play in the evolution of BMPs. As implementers of forest policy and prescriptions, forest and water managers can assess the effectiveness of BMPs relative to the broader goals of contemporary forest management. Similarly, through their role as implementors of BMPs, managers can assist in BMP evolution to keep BMP design and goals current with contemporary management practices.

Recommendations for Managers to Assist the Evolution of BMPs

- Managers should catalogue individual or agency BMP use, design, and goals at the national level and make this information available to the public;
- Managers should undertake monitoring to measure effectiveness of individual BMPs as well as cumulative effects of BMPs; and
- Managers should coordinate these monitoring results with regional state, federal, or citizens groups to assist in the evolution of BMPs in an adaptive management framework.

Adaptive Management

Adaptive management is an approach to natural resources management that promotes carefully designed management actions, assessment of the impact of these actions, and subsequent policy adjustments. An adaptive management strategy explores means of coupling natural and social systems in mutually beneficial ways. Adaptive management recognizes that natural and social systems are not static; they evolve in ways that are often unpredictable over both time and space. In addition to flux in natural systems, adaptive management assumes that human systems change and human interventions induce subsequent ecological adjustments (NRC, 2002). Adaptive management seeks to narrow differences among stakeholders by encouraging them to implement new approaches that will allow people to live with and profit from natural ecosystem variability at socially acceptable levels of risk (Light et al., 1989).

Adaptive management can help managers learn to protect land and water resources, using experiments, monitoring, and modeling. In forest watershed management, adaptive management means the design of forest management actions based on consensus among stakeholders, monitoring of experiment outcomes, and redesign of forest management practices based on this learning. Monitoring and modeling in the context of adaptive management could permit assessment of cumulative watershed effects that encompass complex interactions of water flow, quality, and sediment between headwater catchments and downstream areas (MacDonald and Coe, 2007).

There are limitations to the adaptive management approach. Some management-induced responses may be difficult to detect, particularly at large scales, because they may be small in relation to natural variability or delayed in time after the management action. Adaptive management approaches are also limited in situations where management activities or the resource changes are nearly irreversible.

Despite these limitations, adaptive management offers a framework for managers to work effectively with scientists and stakeholders. Managers can participate in adaptive management by (1) engaging in research-manager partnerships to identify properties of watersheds that should be monitored, (2) conducting the monitoring of these properties, and (3) interpreting and communicating results of monitoring. Modeling is often needed to interpret monitoring results, particularly at larger scales where various land use changes combine to yield an integrated response (see recommendation for modeling by scientists, above) or to incorporate monitoring results into an adaptive management design.

Recommendations for Managers in Adaptive Management

- Managers should design adaptive management approaches for forested watersheds that coordinate management, research, monitoring, and modeling efforts;
- Managers should work with scientists to formulate adaptive management experiments and strategies that assess the effectiveness of current forest management practices relative to contemporary issues at both the local project scale and the large watershed scale; and
- Managers should establish rigorous, consistent monitoring programs, analyze the data collected, and use these data to adapt their management.
- Because there are unavoidable lags in ecological, environmental, and social responses to management practices, it is essential that agencies commit to the adaptive management process for multiple decades.

RECOMMENDATIONS FOR CITIZENS AND COMMUNITIES

Over the past decades, watersheds that were once mostly forested and under single ownership have become fragmented forest patches within a mosaic of land uses and ownership (Chapters 2 and 4). This mosaic obscures the direct effects of forest management and disturbance on hydrology, and makes it difficult to ascribe cumulative watershed effects to specific forest management actions (see Chapter 4). CWEs are sometimes most easily understood during extreme events (see Box 5-3) in large watersheds. Such events remind communities that forest management and disturbance effects on hydrologic processes pervade all ecosystems, including human-dominated ones.

Water researchers and policy makers have long recognized the benefit of organizing land and water management around hydrologic systems (WWPRAC, 1998; Brooks et al., 2003) and have promoted an integrated approach to watershed management (NRC, 1999). Integrated watershed management is an approach that can help identify water movement from headwaters through various land uses to sustainable water supply and quality; it can also provide a means to appraise or manage forest effects on water. The complexity of the water resource and the variety of institutions that govern or oversee it present major challenges. Despite the obvious physical connections, surface water and groundwater are often studied, owned, and regulated as separate resources. Efforts are under way in many states to connect surface and groundwater management, particularly in states facing impending water scarcity, such as Arizona, California, and Colorado (Blomquist et al., 2004).

Although the research community and related entities have recognized the benefits of integrated watershed management, this management has been and largely remains fragmented within and across watershed boundaries. Increasing specialization and pressure from local community groups can create an impetus to coordinate watershed efforts and use integrated watershed management in this coordination.

Watershed Councils

Watersheds are natural units fractured by ownership and land use. Cumulative watershed effects, changes in land ownership, changing population and development patterns, and water supply concerns have spurred local efforts to reconnect watershed hydrology and land use from the community and grass-roots level. New community-level watershed councils and forest groups are proactive in watershed-based and locally driven restoration and management in some areas.

Collaborative watershed institutions, such as watershed management partnerships, councils, or districts, can facilitate the cooperation and collaboration necessary to achieve integrated watershed management. Watershed partnerships, councils, and districts have proliferated across the country in recent decades (Kenney et al., 2000; Brooks et al., 2003; Sabatier et al., 2005; Gregersen et al., 2007). These locally led groups share some common attributes in that most (1) use watershed boundaries to define their jurisdictions; (2) involve a wide variety of agencies and stakeholders from all levels of government and society and treat all participants as equals; (3) negotiate face-to-face to solve problems using collaborative methods; (4) seek mutually beneficial and consensus solutions to watershed management

> **BOX 5-3**
> **Extreme Storms of December 2007 in Washington and Oregon: Cumulative Watershed Effects and Monitoring**
>
> In December 2007, Interstate 5, the major north-south transportation artery for the west coast of the United States was closed for 10 days by flooding that inundated homes and stores and stopped most truck-based transportation (see photo). The flooding was the result of an extreme storm event that produced hurricane force winds along the Oregon coast and delivered more than 35 cm of precipitation in 24 hours to the headwaters of the Chehalis River in central western Washington. In this area, about 30 river gages recorded peak flows that ranked in the top five events, and 10 gages recorded all-time highs. According to the U.S. Geological Survey, this was a 100- to 500-year flood event, based on reconstructed flood magnitudes. Flooding around I-5 was exacerbated by debris flows that carried a mixture of sediment, large wood, and water down the south fork of the Chehalis River (see photo). The landslides that generated the debris flows originated in steep lands owned by the Weyerhauser Corporation along Stillman Creek in the headwaters of the south fork of the Chehalis River. These lands had been clear-cut three and a half years before. Events of this type are relatively common in the Pacific Northwest, where timber is an important industry, slopes are steep and prone to landsliding, and intense storms can deliver large amounts of rainfall. The complex circumstances in this case illustrate the challenges of identifying and disentangling the direct effects of forest management on hydrologic processes from the indirect and interacting effects of storm size, precipitation, and wind speeds.

problems; and (5) build common understanding through extensive, collective fact finding to develop shared understanding of problems and opportunities.

Working across a variety of political, technical, social, and economic boundaries poses continuing challenges, but a watershed council's focus on a specific place gives it a context and sense of shared goals that can begin to bridge jurisdictional and disciplinary boundaries. Aplet et al. (1993) recommended community-based institutions to facilitate coordination among various owners to achieve disparate goals within common ecological and social settings. Citizen groups can participate in this community-based coordination by using existing regulations to provide input (public comment) about timber harvest plans and practices affecting land use, water quality, endangered species, fire prevention, wetlands, and forest chemicals, on both private and public lands.

Photo credit: *http://blog.oregonlive.com/breakingnews/2007/12/large_chehalis-flood-01.jpg*. Reprinted with permission from Oregon Live LLC. Copyright 2007 by Oregonian Publishing Company. SOURCES: Reiter (2008); Seattle Times (2007).

Local collaborations are venues in which the effects of forest management are addressed at the watershed scale with buy-in from owners, providing a highly effective forum for cooperation across ownerships. These new community groups can provide a basis for integrated watershed management that involves a variety of existing institutions responsible for water supply and land management. Community groups would not replace the technical expertise represented by agencies, but they could help initiate and direct management and restoration actions within watersheds.

Recommendations to Advance Community Watershed Groups

- Citizens should request the USFS, U.S. Bureau of Reclamation, and other federal agencies to provide technical expertise in forest or water resources management to community watershed councils and institutions that focus on local watershed management or restoration. Federal agencies should be authorized and funded to provide such technical assistance.
- Citizens should request—and federal and state governments should provide—financial and information resource support for these organizations to ensure continued operation.
- Citizens should participate in management and restoration actions within watersheds.

Community Engagement with Industry and Federal Agencies

"Green certification" for sustainable forest management provides another incentive for forest managers to address water quantity and quality issues. Certification is increasingly necessary for large forest landowners to maintain public acceptance and market access for wood products. For example, the Forest Stewardship Council's forestry management principles include, "Conserve biological diversity, water resources, soils, and unique and fragile ecosystems and landscapes, maintaining the ecological functions and integrity of the forest" (Washburn and Miller, 2003). Some certification systems explicitly require some level of research support to improve forest management as part of their certification requirements, and all require monitoring of forest practices. There is continued controversy about the effects of forest certification on biological diversity (Ghazoul, 2001) and whether forest certification is producing social change (Taylor, 2005). No research has examined how forest certification affects water quantity and quality.

Several federal programs now provide opportunities for community groups to influence local watershed management. The 2003 Healthy Forest Restoration Act (HFRA) aims to accelerate hazardous fuel reduction and forest restoration projects on federal lands at risk of fire or insect and disease epidemics. If communities develop their own Community Wildfire Protection Plan across public and private ownerships, the community receives funding priority under the National Fire Plan, and the Forest Service and the Bureau of Land Management (BLM) can expedite the fuel treatments through alternative environmental compliance options. HFRA also contains a watershed assistance provision (Title III) allowing the Department of Agriculture to provide technical, financial, and related assistance on non-federal forested land and potentially forested land.

In 2003, federal laws were revised to allow so-called stewardship contracting with communities, the private sector, and others. The USFS and BLM may now enter into contracts with local organizations for up to 10 years to improve forest and rangeland health. Stewardship contracts focus on producing desirable results on the ground that improve forest and rangeland health and provide benefits to communities. For example, stewardship contracting allows private organizations or businesses to do thinning and remove small trees and undergrowth; as partial payment, they are able to keep part of what they remove (*http://www.forestsandrangelands.gov/stewardship/index.shtml#projects*). Although many contracts focus on reducing fire risk, example projects also include objectives specifying effects on water from forests, such as "to reduce fuel levels and improve water quality consistent with healthy forest and watershed conditions," or "to improve wildlife habitat, restore sagebrush-steppe habitat, and improve water flow through drainages," or "removal of ponderosa pine and juniper, both of which invaded the riparian corridors and related drainages and are currently out-competing native riparian species."

Recommendations for Community Engagement with Industry and Federal Agencies

- Communities should use public comment opportunities to provide information about the effects of forest management plans on local communities.
- Communities should develop and promote forest certification programs that consider the effects of forest management on water resources.
- Communities should engage in forest stewardship contracting with federal agencies and promote scientifically rigorous monitoring of these forest stewardship contracting projects to determine their effects on water quantity and quality.

MOVING FORWARD: FOREST HYDROLOGY SCIENCE AND MANAGEMENT IN THE 21ST CENTURY

This review and assessment of the state of forest hydrology knowledge at the beginning of the twenty-first century provides major findings regarding the current understanding of forest hydrology as well as information gaps and research needs to advance forest hydrology from principles to predictions for management. It also offers recommendations to meet those research needs for forest hydrology science and management (Table 5-1).

Forest hydrology science has produced a solid understanding of the general principles and basic processes of how water is connected to and moves through forests. The current forest landscape is dynamic due to changing demographics, climate patterns, land use and ownership, and the increased demand for water. Forest science and management are adapting as the land uses and land ownership within forested watersheds become more heterogeneous, changes in climate and its effects are becoming more evident, and it is easier to visualize cumulative watershed effects over larger spatial scales and longer periods of time. The strong foundation of general principles and basic processes in forest hydrology can be applied to meet research needs and fill information gaps over the coming decades.

TABLE 5-1 Current Understanding, Research Needs, and Recommendations for Sustaining Water Supplies from Forests

	Current Understanding	Information Gaps and Research Needs	Recommended Actions
Science	The body of forest hydrology science derives from almost 100 years of studies at small spatial and time scales Forest hydrology science has established general principles that are understood with a high degree of certainty describing direct hydrologic effects of forest management and disturbance Effects can be understood through changes in • Forest structure • Magnitudes, rates, and flowpaths • Erosion, nutrient cycling, and soil chemistry Reduced forest cover results in increased water yield that is • Generally short-lived • Greatest during times of water excess rather than water scarcity • Small or undetectable in water-scarce areas • May be associated with a decline in water quality	Hydrologic effects of past management, such as fire suppression, clear-cutting, roads Ways to quantify hydrologic responses at larger spatial and temporal scales Ways to scale up findings from small spatial and short time scales to larger spatial and longer time scales Use general principles to predict indirect hydrologic responses to changes in forest landscapes and interacting responses to forest management and disturbance	Enhance, maintain, and reestablish abandoned small watershed studies Combine existing data from the large body of small watershed studies and analyze them for large-scale trends as a meta-experiment Use new technologies, including sensor networks and remote sensing, to improve understanding of forest hydrology in changing landscapes Engage in adaptive management to help managers and community groups design monitoring strategies, develop and test models, and conduct studies relevant to management
Management	Forests in the United States are managed for a wide range of goals and objectives: timber harvesting,	Assessment of BMP effectiveness Principles and practices of adaptive management	Advance BMP evolution by rigorously assessing and developing new BMPs and

Scale			
	road networks and road construction, high-severity wildfires, and exurban sprawl modify forest hydrology Forest management practices are evolving in response to environmental change, social and economic forces, and technological developments BMPs are used to mitigate impacts on water resources from forest management activities		measuring their effectiveness At the federal level, provide sustained support for adaptive management activities, enabling managers to partner with scientists to design and implement monitoring, develop and test models, and conduct studies relevant to management issues Increase role of agency technical expertise in watershed councils
Community	Integrated watershed management is a viable vehicle for both community groups and state and federal agencies to help manage water and forest resources at the community scale Citizens groups can influence local and integrated watershed management Community watershed groups benefit from state and federal agency technical expertise Existing laws can be used to strengthen the standing and influence of watershed councils New laws offer increased opportunities for community involvement	How watershed councils and their stakeholders view and utilize forest hydrology science and scientific expertise from federal agencies How industry-sponsored green certification and federal forest stewardship contracts affect water quantity and quality from forests	Use watershed councils to meet multiple goals of integrated watershed management at the community level Expand the number and influence of watershed councils. Engage in adaptive management with scientists and managers

problems; and (5) build common understanding through extensive, collective fact finding to develop shared understanding of problems and opportunities.

Working across a variety of political, technical, social, and economic boundaries poses continuing challenges, but a watershed council's focus on a specific place gives it a context and sense of shared goals that can begin to bridge jurisdictional and disciplinary boundaries. Aplet et al. (1993) recommended community-based institutions to facilitate coordination among various owners to achieve disparate goals within common ecological and social settings. Citizen groups can participate in this community-based coordination by using existing regulations to provide input (public comment) about timber harvest plans and practices affecting land use, water quality, endangered species, fire prevention, wetlands, and forest chemicals, on both private and public lands.

References

Aber, J.D. 1992. Nitrogen cycling and nitrogen saturation in temperate forest ecosystems. Trends in Ecology and Evolution 7(7):220-224.

Aber, J.D., S.V. Ollinger, C.A. Federer, P.B. Reich, M.L. Goulden, D.W. Kicklighter, J.M. Melillo, and R.G. Lathrop, Jr. 1995. Predicting the effects of climate change on water yield and forest production in the northeastern United States. Climate Research 5: 207-222.

Adams, M.B., P.J. Edwards, F. Wood, and J.N. Kochenderfer. 1993. Artificial watershed acidification on the Fernow Experimental Forest, USA. Journal of Hydrology 150:505-519.

Alig, R. 2006. Society's choices: land use changes, forest fragmentation, and conservation. U.S. Forest Service, Pacific Northwest Research Station, Findings 88.

Alig, R., and A. Plantinga. 2004. Future forest land area: Impacts from population growth and other factors affecting land values. Journal of Forestry 102(8):19-24

Alig, R.J., A.J. Plantinga, S. Ahn, and J.D. Kline. 2003. Land use changes involving forestry in the US: 1952 to 1997, with projections to 2050. Gen. Tech. Rep. PNW-GTR-587. Portland, OR: U.S. Department of Agriculture, Forest Service, Pacific Northwest Research Station.

Alila, Y., and J. Beckers. 2001. Using numerical modeling to address hydrologic forest management issues in British Columbia, Hydrological Processes 15(SI): 3371-3387.

Anderson, C. W. 2002. Ecological effects on streams from forest fertilization-Literature review and conceptual framework for future study in the western Cascades. U.S. Geological Survey Water Resources Investigations Report 01-4047.

Anderson, H.W. 1974. Sediment deposition in reservoirs associated with rural roads, forest fires, and catchment attributes. Pp. 13:87-95 In Proceedings symposium on man's effect on erosion and sedimentation, UNESCO, 9-12 September, Paris, IAHS.

Anderson, H.W., M.D. Hoover, and K.G. Reinhart. 1976. Forests and Water: Effects of Forest Management on Floods, Sedimentation, and Water Supply. General Technical Report PSW-018. Berkeley, CA: U.S. Department of Agriculture, Forest Service, Pacific Southwest Forest and Range Experiment Station.

Aplet, G. H., N. Johnson, J.T. Olson, and V.A. Sample. 1993. Prospects for a Sustainable Future. Defining Sustainable Forestry. Washington, DC: Island Press.

Armour, C., D. Duff, and W. Elmore. 1994. The effects of livestock grazing on western riparian and stream ecosystems. Fisheries 19(9):9-12.

Arno, S. F., and J. K. Brown. 1991. Overcoming the paradox in managing wildland fire. Western Wildlands 17:40-46.

Arthur, M.A., G.B. Coltharp, and D.L. Brown. 1998. Effects of best management practices on forest streamwater quality in eastern Kentucky. Journal of the American Water Resources Association 34 (3):481–495.

Auch, R., J. Taylor, and W. Acevedo. 2004. Urban growth in American cities: Glimpses of U.S. urbanization. U.S. Geological Survey Circular 1252.

Aust, W.M. and C.R. Blinn. 2004. Forestry best management practices for timber harvesting and site preparation in the Eastern United States: An overview of water quality and productivity research during the past 20 years (1982-2002). Water, Air, and Soil Pollution: Focus 4:5-36.

Austin, S. A. 1999. Streamflow response to forest management: a meta-analysis using published data and flow duration curves. M. S. Thesis, Department of Earth Resources, Colorado State University.

Baker, J.P., and D.H. Landers. 2003. Alternative futures analysis for the Willamette River- basin, Oregon. Ecological Applications 14(2):311–400.

Baker, M.B., Jr., and P.F. Ffolliott. 1998. Multiple resource evaluations on the Beaver Creek watershed: An annotated bibliography of 40 years of investigations. USDA Forest Service, General Technical Report RM-GTR-13.

Baker, S., A. Richardson, and L. Barmuta. 2004. Site effects outweigh riparian influences on ground-dwelling beetles adjacent to first order streams in wet eucalypt forest. Biodiversity and Conservation 16:1999-2014.

Baker, W.L. 1992. Effects of settlement and fire suppression on landscape structure. Ecology 73(5):1879-1887

Baldigo, B.P., and G.B. Lawrence. 2001. Effects of stream acidification and habitat on fish populations of a North American river. Aquatic Science 63:196–222.

Barnett, T., R. Malone, W. Pennell, D. Stammer, B. Semtner, and W. Washington. 2004. The effects of climate change on water resources in the West: Introduction and overview. Climatic Change 62(1-3):1-11.

Barnett, T .P., D.W. Pierce, G.H. Hidalgo, C.Bonfils, D. B. Santer, T. Das, G. Bala, A. Wood, T. Nozawa, A. Mirin, D.R. Cayan, and M.D. Dettinger. 2008. Human-induced changes in the hydrology of the western United States. Science 319: 1080-1083.

Barten, P.K., and C.E. Ernst. 2004. Land conservation and watershed management for source protection. Journal of the American Water Works Association 96(4):121-135.

Barton, D.R., W.D. Taylor, and R.M Biette. 1985. Dimensions of riparian buffer strips required to maintain trout habitat in southern Ontario streams. North American Journal of Fisheries Management 5:364–378.

Bates, C.G., and A.J. Henry. 1928. Forest and stream-flow experiment at Wagon Wheel Gap, Colorado. U.S. Weather Bureau Monthly Weather Review, Supplement No. 30.

Bates, D., K. Willis, F. Swanson, J.R. Glasmann, D. Halemeier, and H. Wujcik. 1998. North Santiam River turbidity study, 1996–1997. Berkeley, CA: Watershed Management Council, University of California. Watershed Management Council Networker 1998:1-17.

Bebi, P., D. Kulakowski, and T. T. Veblen. 2003. Interactions between fire and spruce beetles in a subalpine Rocky Mountain forest landscape. Ecology 84:362-371.

Bebi, P., D. Kulakowski, and T.T. Veblen. 2003. Interactions between fire and spruce beetles in a subalpine Rocky Mountain forest landscape. Ecology 84(2):362–371.

Beebe, G. S., and P. H. Omi. 1993. Wildland burning. Journal of Forestry 91(9):19-24.

Bell, J. W. 2000. The National Forest road system: A policy issue for the 21st century. In: Flag, M., and D. Revert (Eds.) Watershed management 2000: Proceedings. (CD-ROM)

Belsky, A.J., and D.M. Blumenthal. 1997. Effects of livestock grazing on stand dynamics and soils in upland forests of the interior west. Conservation Biology 11(2):315-327.

Benavides-Solorio, J.D., and L.H. MacDonald. 2005. Measurement and prediction of post-fire erosion at the hillslope scale, Colorado Front Range. International Journal of Wildland Fire 14:457-474.

Benda, L. D. Miller, K. Anras, P. Bigelow, G. Reeves, and D. Michael 2007. NetMap: A new tool in support of watershed science and resource management. Forest Science 53(2):206-219.

Bergeron, Y., and A. Leduc. 1998. Relationships between Change in fire frequency and mortality due to spruce budworm outbreak in the southeastern Canadian boreal forest. Journal of Vegetation Science 9(4): 493-500.

Bergeron, Y., and P.-R. Dansereau. 1993. Predicting the composition of Canadian southern boreal forest in different fire cycles. Journal of Vegetation Science 4(6):827-832.

Berndt, H.W. 1971. Early Effects of Forest Fire on Streamflow Characteristics. USDA Forest Service, Pacific Northwest Research Station, Research Note PNW-148.

Beschta, R.L., and R.L. Taylor. 1988. Stream temperature increases and land use in a forested Oregon watershed. Journal of the American Water Resources Association 24 (1):19–25.

Beschta, R.L., M.R. Pyles, A.E. Skaugset, and C.G. Surfleet. 2000. Peakflow responses to forest practices in the western cascades of Oregon, USA. Journal of Hydrology 233: 102-120.

Bethlahmy, N. 1974. More streamflow afer a bark beetle epidemic. Journal of Hydrology 23:185-189.

Bilby, R.E. 1985. Contributions of road surface sediment to a western Washington stream. Forest Science 31(4):827-838.

Binkley, D. 2001. Patterns and processes of variation in nitrogen and phosphorus concentrations in forested streams, Technical Bulletin 836. Research Triangle Park, NC: National Council for Air and Stream Improvement.

Binkley, D., and T. C. Brown. 1993a. Forest practices as nonpoint sources of pollution in North America. Water Resources Bulletin 29(5):729–740.

Binkley, D., and T.C. Brown. 1993b. Management Impacts on Water Quality of Forests and Rangelands. Gen. Tech. Rep. RM–239. Fort Collins, CO: U.S. Department of Agriculture, Forest Service, Rocky Mountain Forest and Range Experiment Station.

Binkley, D., D.H. Burnham, and H.L. Allen. 1999. Water quality impacts of forest fertilization with nitrogen and phosphorus. Forest Ecology and Management 121: 191–213.

Binkley, D., G.G. Ice, J. Kaye, and C.A. Williams. 2004. Nitrogen and phosphorus concentrations in forest streams of the United States. Journal of the American Water Resources Association 40(5):1277-1291.

Bisson, P.A., T.P. Quinn, G.H. Reeves, and S.V. Gregory. 1992. Best management practices, cumulative effects, and long-term trends in fish abundance in Pacific Northwest river systems. Pp. 189-232 in R. Naiman (ed.) Watershed management: balancing sustainability and environmental change. New York, Springer.

Blinn, C.R., and M.A Kilgore. 2001. Riparian Management Practices: A Summary of State Guidelines. Journal of Forestry 99(8): 11-17(7).

Blomquist, W., E. Schlager, and T. Heikkila. 2004. Common waters, diverging streams: linking institutions and water management in Arizona, California and Colorado. Resources for the Future, Washington, D.C.

Bolin, S.B., and T.J. Ward. 1987. Recovery of a New Mexico drainage basin from a forest fire. In Proceedings of the symposium on forest hydrology and watershed management. IAHS 167:191-198.

Boomer, K.E., D.E. Weller, and T.E. Jordan. 2008. Empirical models based on the Universal Soil Loss Equation fail to predict sediment discharges from Chesapeake Bay catchments. Journal of Environmental Quality37(1):79-89.

Bosch, J. M., and J. D. Hewlett. 1982. A review of catchment studies to determine the effect of vegetative changes on water yield and evapotranspiration. Journal of Hydrology 55:3-23.

Bowling, L.C. and D.P. Lettenmaier 2001. The effects of forest roads and harvest on catchment hydrology in a mountainous maritime environment. Water Science and Application 2: 145-164.

Briggs, R.D., J. Cormier, and A. Kimball. 1998. Compliance with Forestry Best Management Practices in Maine. Northern Journal of Applied Forestry 15(2): 57-68(12).

Broadmeadow, S., and T.R. Nisbet. 2004. The effects of forest management on the freshwater environment: a literature review of best management practice. Hydrology and Earth System Science 8(3): 286-305.

Brockway, D.G., and C.E. Lewis. 1997. Long-term effects of dormant-season prescribed fire on plant community diversity, structure and productivity in a longleaf pine-wiregrass ecosystem. Forest Ecology and Management 96: 167-183.

Brooks, K.N., P.F. Ffolliott, H.M. Gregersen and L.F. DeBano. 2003. Hydrology and the management of watersheds. 3ed. Iowa State Press, Ames.

Brown, A.E., L. Zhang, T.A. McMahon, A.W. Western, and R.A. Vertessy. 2005. A review of paired catchment studies for determining changes in water yield resulting from alterations in vegetation. Journal of Hydrology 310: 28-61.

Brown, J.H., T.J. Valone, and C.G. Curtin. 1997. Reorganization of an arid ecosystem in response to recent climate change. Proceedings of the National Academy of Science 94(18): 9729–9733.

Brown, T.C., and D. Binkley. 1994. Effect of Management on Water Quality in North American Forests. General Technical Report RM-248. Washington, DC: USDA Forest Service,

Brown, T.C., D. Brown, D. Binkley. 1993. Laws and programs for controlling nonpoint-source pollution in forest areas. Water Resources Bulletin 29, 1-13.

Bue, C.D., M.T. Wilson, and E.L. Peck. 1955. Discussion of The effect on streamflow of the killing of spruce and pine by the Engelmann spruce beetle. Transactions of the American Geophysical Union 36(6):1087-1089.

Buhl, K.J., and S.J. Hamilton. 1998. Acute toxicity of fire-retardant and foam-suppressant chemicals to early life stages of chinook salmon (*Onchorhnchus tshawytscha*): Environmental Toxicology and Chemistry 17(8):1589-1599.

Buhl, K.J., and S.J. Hamilton. 2000. Acute toxicity of fire-control chemicals, nitrogenous chemicals, and surfactants to rainbow trout: Transactions of the American Fisheries Society 129:408-418.

Bytnerowicz, A., M. J. Arbaugh, and S. L. Schilling (technical coordinators). 1998. Proceedings of the international symposium on air pollution and climate change effects on forest ecosystems. USDA Forest Service, General Technical Report PSW-GTR-166.

Caissie, D., S. Jolicoeur, M. Bouchard, and E. Poncet. 2002. Comparison of streamflow between pre and post timber harvesting in Catamaran Brook (Canada). Journal of Hydrology 258(1-4):232-248.

Callaway and Aschehoug. 2000.

Campbell, I.C., and T.J. Doeg. 1989. Impact of timber harvesting and production on streams: A review. Australian Journal of Marine and Freshwater Research 40(5): 519-539

Campbell, R.E., M.B. Baker, Jr., P.F. Ffolliott, F.R. Larson, and C.C. Avery. 1977. Wildfire Effects on a Ponderosa Pine Ecosystem: An Arizona Case Study. Fort Collins, CO: USDA Forest Service, Rocky Mountain Research Station, Research Paper RM-191, .

Cappiella, K., T. Schueler, and T. Wright. 2005. Urban Watershed Forestry Manual, Part 1: Methods for Increasing Forest Cover in a Watershed. Newtown Square, PA: USDA Forest Service NA State and Private Forestry Publication NA-TP-04-05.

Carpenter, S.R. 1998. The need for large-scale experiments to assess and predict the response of ecosystems to perturbation. P. 287-312. In Pace, M.L., and P.M. Groffman (eds.) Successes, Limitations, and Frontiers in Ecosystem Science. New York: Springer-Verlag.

Cerda, A. 1998. Relationships between climate and soil hydrological and erosional characteristics along climatic gradients in Mediterrainean limestone areas. Geomorphology 25: 123-134.

Chang, M. 2003. Forest Hydrology: An Introduction to Water and Forests. Boca Raton, FL: CRC Press.

Christensen, N.L., A. Bartuska, J.H. Brown, S. Carpenter, C. D'Antonio, R. Francis, J.F. Franklin, J.A. MacMahon, R.F. Noss, D.J. Parsons, C.H. Peterson, M.G. Turner, and R.G. Woodmansee. 1996. The scientific basis for ecosystem management. Ecological Applications 6:665-691.

Christensen, N.S., A.W. Wood, N. Voisin, D.P. Lettenmaier, and R.N. Palmer. 2004. The effects of climate change on the hydrology and water resources of the Colorado River Basin. Climatic Change 62(1-3):337-363.

Clark, J.S. 1990. Twentieth-century climate change, fire suppression, and forest production and decomposition in northwestern Minnesota. Canadian Journal of Forest Research 20(2):219–232.

Coe, D. 2006. Sediment production and delivery from forest roads in the Sierra Nevada, California. M.S. Thesis, Department of Earth Resources, Colorado State University.

Cortner, H.J., P.D. Gardner, and J.G. Taylor. 1990. Fire hazards at the urban wildland interface—What the public expects. Environmental Management 14:57-62.

Covington, W.W., and M.M. Moore. 1994. Southwestern Ponderosa forest structure: Changes since Euro-American settlement. Journal of Forestry 92(1):39-47(9).

Cuo, L., T. W. Giambelluca, A. D. Ziegler, and M. A. Nullet. 2006. Use of the distributed hydrology soil vegetation model to study road effects on hydrological processes in Pang Khum Experimental Watershed, northern Thailand. Forest Ecology and Management 224: 81-94.

Czech, B., and P. F. Ffolliott. 1996. The let-burn policy: Implications in the Madrean Province of the southwestern United States. Pp. 223-225 In: Ffolliott, P. F., L. F. DeBano, M. B. Baker, Jr., G. J. Gottfried, G. Solis-Garza, C. B. Edminster, D. G. Neary, L. S. Allen, and R. H. Hamre (technical coordinators) Effects of fire on Madrean Province ecosystems. USDA Forest Service, General Technical Report RM-GTR-289.

Dahm, C.N., J.R. Cleverly, J.E. Allred Coonrod, J.R Thibault, D.E. McDonnell, and D.J. Gilroy. 2002. Evapotranspiration at the land/water interface in a semi-arid drainage basin. Freshwater Biology 47 (4):831–843.

Dale, V.H., L.A. Joyce, S. McNulty, R.P. Neilson, M.P. Ayres, M.D. Flannagan, P.J. Hanson, L.C. Irland, A.E. Lugo, C.J. Peterson, D. Simberloff, F.J. Swanson, B.J. Stocks, and B.M. Wotton. 2001. Climate change and forest disturbances. BioScience 51(9) : 723-734.

Darveau, M., P. Beauchesne, L. Belanger, J. Huot, and P. Larue. 1995. Riparian forest strips as habitat for breeding birds in Boreal forests. Journal of Wildlife Management 59(1):67-78.

Davis, L. S., K. N. Johnson, P. Bettinger, and T. Howard. 2000. Forest Management. New York: McGraw-Hill.

Davis, M.B., and R.G. Shaw. 2001. Range shifts and adaptive responses to quaternary climate change. Science 292(5517):673 - 679

de la Crétaz, A.L., and P.K. Barten. 2007. Land Use Effects on Streamflow and Water Quality in the Northeastern United States. Boca Raton, FL: CRC Press.

DeBano, L.F. 2000. The role of fire and soil heating on water repellency in wildland environments: a review. Journal of Hydrology 231-232:195-206.

DeBano, L.F., D.G. Neary, and P.F. Ffolliott. 2005. Soil physical properties. Pp. 29-52 In Neary, D.G., K.C. Ryan, and L.F. DeBano (eds.) Wildland Fire in Ecosystems: Effects of Fire on Soil and Water. USDA Forest Service General Technical Report RMRS-GTR-42, Volume 4.

DeBano, S.J., and D. E. Wooster. 2004. Insects and other invertebrates: Ecological roles and indicators of riparian and stream health. Pp. 215-236 In Baker, M. B. Ffolliott, P. F., DeBano, L. F. and D. G. Neary (eds.) Riparian Areas of the Southwestern United States: Hydrology, Ecology, and Management. Boca Raton, FL: Lewis Publishers.

DeHayes, D.H., P.G. Shaberg, G.J. Hawley, and G.R. Strimbeck. 1999. Acid rain impacts calcium nutrition and forest health. BioScience 49: 789-800.

Dent, L., and J. Robben. 2000. Oregon Department of Forestry: Aerial pesticide application monitoring–Final report. Forest Practices Monitoring Program technical report 7. Salem, OR: Oregon Department of Forestry.

Dettinger, M., D. Cayan, M. Meyer, and A. Jeton. 2004. Simulated hydrologic responses to climate variations and change in the Merced, Carson, and American River Basins, Sierra Nevada, California, 1900–2099. Climate Change 62:283–317.

Dettinger, M., D. Cayan, M. Meyer, and A. Jeton. 2004. Simulated hydrologic responses to climate variations and change in the Merced, Carson, and American River Basins, Sierra Nevada, California, 1900–2099. Climate Change 62: 283–317.

Dissmeyer, G.E. (ed). 2000. Drinking water from forests and grasslands: a synthesis of the scientific literature. Gen. Tech. Rep. SRS–39. Asheville, NC: U.S. Department of Agriculture, Forest Service, Southern Research Station.

Dissmeyer, G.E. (ed). 2000. Drinking Water from Forests and Grasslands: A Synthesis of the Scientific Literature. Gen. Tech. Rep. SRS–39. Asheville, NC: U.S. Department of Agriculture, Forest Service, Southern Research Station.

Doerr, S.H., and A.D. Thomas. 2000. The role of soil moisture in controlling water repellency: new evidence from forest soils in Portugal. Journal of Hydrology 231-232: 134-147.

Doerr, S.H., R. A. Shakesby, W.H. Blake, C.J. Chafer, G.S. Humphreys, and P.J. Wallbrink. 2006. Effects of differing wildfire severities on soil wettability and implications for hydrological response. Journal of Hydrology 319:295-311.

Dole, D., and E. Niemi. 2003. Future water allocation and in-stream values in the Willamette River Basin: a basin-wide analysis. Ecological Applications 14(2):355–367.

Donato, D.C., J.B. Fontaine, J.L. Campbell, W.D. Robinson, J.B. Kauffman, and B.E. Law. 2006. Post-wildfire logging hinders regeneration and increases fire risk. Science 311:352-352.

Douglas, J.E., and D.H. Van Lear. 1983. Prescribed burning and water quality of ephereral streams in the piedmont of South Carolina. Forest Science 29:181-189.

Douglass, J.E. 1983. The potential for water yield augmentation from forest management in the eastern United States. Journal of the American Water Resources Association 19(3):351-358.

Dunne, T., and L.B. Leopold. 1978. Water in Environmental Planning. San Francisco, CA: W.H. Freeman & Company.

Edwards, P.J., and J.D. Helvey. 1991. Long-term ionic increases from a control Appalachian forested watershed. Journal of Environmental Quality 20:250–255.

Eisenbies, M.H., W.M. Aust, J.A. Burger and M.B. Adams. 2007. Forest operations, extreme flooding events, and considerations for hydrologic modeling in the Appalachians–a review. Forest Ecology and Management 242:77-98.

Ewing, R. 1996. Postfire suspended sediment from Yellowstone National Park, Wyoming. Water Resources Bulletin 32:605-627.

Falk, D. A., and T. W. Swetnam. 2003. Scaling rules and probability models for surface fire regimes in ponderosa pine forests. Pp. 301-317 In P. N. Omi, and L. A. Joyce (eds.) Fire Ecology, Fuel Treatments, and Ecological Restoration. USDA Forest Service. Proceedings RMRS-P-29.

FAO (Food and Agricultural Organization). 2005. Forests and floods – drowning in fiction or thriving on facts? RAP Publication. Forest Perspectives. Bangkok, Thailand: United Nations.

Fenn, M.E., M.A. Poth, J.D. Aber, and others. 1998. Nitrogen excess in North American ecosystems: predisposing factors, ecosystem responses, and management strategies. Ecological Applications 8: 706–733.

Ffolliott, P. F., L. A. Bojorquez-Tapia, and M. Hernandez-Nerves. 2001. Natural Resources Management Practices: A Primer. Ames, IA: Iowa State University Press.

Ffolliott, P.F. and D.G. Neary. 2003. Impacts of a historical wildfire on hydrologic processes: A case study in Arizona. AWRA 2003 International Congress. (CD-Rom). Middleburg, VA: American Water Resources Association,

Flather, C. H., L. A. Joyce, and C. A. Bloomgarden. 1994. Species Endangerment Patterns in the United States. USDA Forest Service, General Technical Report RM-241.

Fleming, R. A., J. Candau, and R. S. McAlpine. 2002. Landscape-scale analysis of interactions between insect defoliation and forest fire in central Canada. Climate Change 55:251-272.

Flum, T., and S.C. Nodvin. 1995. Factors affecting streamwater chemistry in the Great Smoky Mountains, U.S.A. Water, Air and Soil Pollution 85: 1707–1712.

Foster, D. and J. Aber. 2004. Forests in Time: The Environmental Consequences of 1,000 Years of Change in New England. New Haven: Yale University Press.

Foster, D., D. Kittredge, B. Donahue, G. Motzkin, et al. 2005. Wildlands and Woodlands A Vision for the Forests of Massachusetts. Cambridge, MA: Conservation Research Foundation, the Fine Family Foundation, Friends of the Harvard Forest and Sweet Water Trust.

Foster, D.R., and E.R. Boose. 1992. Patterns of forest damage resulting from catastrophic wind in central New England, USA. The Journal of Ecology 80(1):79-98.

Fried, J.S., M.S. Torn, and E. Mills. 2004. The impact of climate change on wildfire severity: A regional forecast for Northern California. Climatic Change 64: 169-191.

GAO (General Accounting Office). 1996. Public Timber: Federal and State Programs Differ Significantly in the Pacific Northwest. Report of the Chairman, Committee on Resources, U.S. House of Representatives. Washington, DC: GAO.

GAO. 1997. Minerals Management: Costs for Onshore Minerals Leasing Programs in Three States. GAO/RCED-96-108. Washington, DC: GAO.

GAO. 1998. Forest Service: Lack of Financial and Performance Accountability has Resulted in Inefficiency and Waste. GAO/T-RCED/ALMD-98-135. Washington, DC: GAO.

Ghazoul, J. 2001. Barriers to biodiversity conservation in forest certification. Conservation Biology 15(2): 315-317.

Goetz, S.J., C.A. Jantz, S.D. Prince, A.J. Smith, D. Varlyguin, and R.K. Wright. 2004. Integrated analysis of ecosystem interactions with land use change: The Chesapeake Bay Watershed. Ecosystems and Land Use Change, American Geophysical Union Geophysical Monograph Series 153.

Gottfried, G.J., and L.F. DeBano. 1990. Streamflow and water quality response to preharvest prescribed burning in and undisturbed ponderosa pine watershed. Pp. 222-231 In Krammes, J.S. (ed.) Effects of Fire Management on Southwestern Natural Resources. General Technical Report RMRS-GTR-191. Fort Collins, CO: US Forest Service, Rocky Mountain Research Station.

Graham, R.T., S. McCaffery, and T.B. Jain. 2004. Science basis for changing forest structure to modify wildfire behavior and severity. Gen. Tech. Rep. RMRS-GTR-120. Fort Collins, CO: USDA Forest Service Rocky Mountain Research Station.

Grant, G.E., and A.L. Wolff. 1991. Long-term patterns of sediment transport after timber harvest, Western cascade mountains, Oregon, USA. Sediment and Stream Water Quality in a Changing Environment: Trends and Explanation. Proceedings of the Vienna Symposium, August 1991. IAHS Publ. No. 203.

Grant, G.E., and F. Swanson. 1991. Cumulative effects of forest practices. Forest Perspectives: New Directions in Forest Management 1(4):9-11.

Grant, G.E., S.L. Lewis, F.J. Swanson, J.H. Cissel, and J.J. McDonnell. 2008. Effects of Forest Practices on Peak Flows and Consequent Channel Response: A State-of-Science report for Western Oregon and Washington. General Technical Report PNW-GTR-760. Portland, OR: U.S. Department of Agriculture, Forest Service, Pacific Northwest Research Station.

Gray, A. 2007. Distribution and abundance of invasive plants in Pacific Northwest forests. Pp 143-148 In Harrington, T.B. and S.H. Reichard (tech. eds.). Meeting the Challenge: Invasive Plants in Pacific Northwestern Ecosystems. Gen. Tech. Rep. PNW-GTR-694. Portland, OR: U.S. Department of Agriculture, Forest Service, Pacific Northwest Research Station. Available online http://www.fs.fed.us/pnw/pubs/pnw_gtr694.pdf Accessed May 2, 2008.

Gregersen, H.M., P.F. Ffolliott, and K.N. Brooks. 2007. Integrated watershed management: connecting people to their land and water. Wallingford, Oxfordshire, UK: CAB International.

Guertin, D.P., S.N. Miller, and D.C. Goodrich. 2000. Emerging tools and technologies in watershed management. Pp. 194-204 In Land stewardship in the 21st Century: The Contributions of Watershed Management. USDA Forest Service Proceedings RMRS-P-13. Tucson, AZ: USDA Forest Service.

Hamlet, A.F., and D.P. Lettenmaier 1999. Effects of climate change on hydrology and water resources in the Columbia River Basin. Journal of the American Water Resources Association 35 (6):1597–1623.

Harr, R.D. 1986. Effects of clearcutting on rain-on-snow runoff in western Oregon: A new look at old studies. Water Resources Research 22(7):1095-1100.

Harr, R.D., R.L. Fredricksen, and J. Rothacher. 1979. Changes in streamflow following timber harvest in southwestern Oregon. USDA Forest Service Research Paper. PNW-249. Portland, OR: Pacific Northwest Forest and Range Experiment Station.

Harr, R.D., W.C. Harper, J.T. Krygier, and F.S. Hsieh. 1975. Changes in storm hydrographs after road building and clear cutting in the Oregon Coast Range. Water Resources Research 11(3):436-444.

Harr, R.D. 1976. Hydrology of small forest streams in western Oregon. General Technical Report GTR-PNW-55. Portland, OR: USDA Forest Service.

Harr, R.D. 1981. Some characteristics and consequences of snowmelt during rainfall in western Oregon. Journal of Hydrology 53:277-304.

Harr, R.D. 1982. Fog drip in the Bull Run municipal watershed, Oregon. Journal of the American Water Resources Association 18(5):785–789.

Harr, R.D. 1983. Potential for augmenting water yield through forest practices in western Washington and western Oregon. Journal of the American Water Resources Association 19 (3):383–393.

Harr, R.D., A. Levno, and R. Mersereau. 1982. Streamflow changes after logging 130-year-old douglas fir in two small watersheds. Water Resources Research 18(3): 637-644.

Harr, R.D., and R.L. Fredriksen. 1988. Water quality after logging small watersheds within the Bull Run watershed. Journal of the American Water Resources Association 24(5):1103–1111.

Hawkins, R.H. 1993. Asymptotic determination of runoff curve numbers from data. Journal of Irrigation and Drainage Engineering 119:334-345.

Hawkins, R.H., and A.V. Khojeimi. 2000. Initial abstraction and loss in the curve number method. Hydrology and Water Resources in Arizona and the Southwest 30:29-35.

Heathcote, I.W. 1998. Integrated watershed management: Principles and practice. New York: John Wiley and Sons.

Helvey J.D., and J.H. Patric. 1965, Canopy and litter interception by hardwoods of eastern United States. Water Resources Research 1: 193-206.

Helvey, J.D. 1971. A summary of rainfall interception by certain confiers of North America. Pp. 103-113 In Monke, E.J. (ed.) Proceedings of the Third International Symposium for Hydrology Professors. Biological Effects in the Hydrological Cycle. West Lafayette, IN: Purdue University.

Helvey, J.D. 1980. Effects of a north-central Washington wildfire on runoff and sediment production. Water Resources Bulletin 16:627-634.

Hibbert, A.R. 1967. Forest treatment effects on water yield. Pp. 527-543 In Sopper, W.E., and H.W. Lull (eds.) International Symposium on Forest Hydrology, New York: Pergamon Press.

Hibbert, A.R. 1983. Water yield improvement potential by vegetation management on western rangelands. Water Resources Bulletin 19(3): 375-381.

Hicke, J.A., J.A. Logan, J. Powell, and D.S. Ojima. 2006. Changing temperatures influence suitability for mountain pine beetle outbreaks in the western US. Journal of Geophysical Research-Biogeosciences 111.

Hicks, B.J., R.L. Beschta, and R.D. Harr. 1991. Long-term changes in streamflow following logging in western Oregon and associated fisheries implications. Water Resources Bulletin 27: 217-226.

Hobbie, J.E., and G.E. Likens. 1973. Output of phosphorus, dissolved organic carbon, and fine particulate carbon from Hubbard Brook watersheds. Limnology and Oceanography 18:734–742.

Hodgkins, G.A., R.W. Dudley, and T.G. Huntington. 2003. Changes in the timing of high river flows in New England over the 20th century. Journal of Hydrology 278: 244–252.

Hornbeck, J.W., C.W. Martin, and C. Eagar. 1997. Summary of water yield experiments at Hubbard Brook Experimental Forest, New Hampshire. Canadian Journal of Forest Research 27(12): 2043–2052.

Hornbeck, J.W., M.B Adams, E.S. Corbett, E.S. Verry, and J.A. Lynch. 1993. Long-term impacts of forest treatments on water yield: A summary for northeastern USA. Journal of Hydrology 150 (2/4):323–344.

Houghton, J.T. (ed.). 1995. Climate Change 1995: The Science of Climate Change. Contribution of Working Group I to the Second Assessment Report of the Intergovernmental Panel on Climate Change. Cambridge, MA: Intergovernmental Panel on Climate Change, Cambridge University Press.

Hoyt, W.G., and H.C. Troxell. 1934. Forests and streamflow. Paper no. 1858, Transactions of the American Society of Civil Engineers. Reston, VA: ASCE.

Huffman, E.L., L.H. MacDonald, and J.D. Stednick. 2001. Strength and persistence of fire-induced soil hydrophobicity under ponderosa and lodgepole pine, Colorado Front Range. Hydrological Processes 15:2877-2892.

Hull Sieg, C., B. G. Phillips, and L. P. Moser. 2003. Exotic invasive plants. Pp. 251-267 In Friederici (ed.) Ecological Restoration of Southwestern Ponderosa Pine Forests. Covelo, CA: Island Press.

Hulse, D.W., A. Branscomb, and S.G. Payne. 2003. Envisionsing alternatives: using citizen guidance to map future land and water use. Ecological Applications 14(2): 325–341.

Hutley, L.B., D. Doley, D.J. Yates, and A. Boonsaner. 1997. Water-balance of an Australian subtropical rain-forest at altitude: The ecology and physiological significance of intercepted cloud and fog. Australian Journal of Botany 45: 311-329.

Ice, G. 2004. History of Innovative Best Management Practice Development and its Role in Addressing Water Quality Limited Waterbodies. Journal of Environmental Engineering 130(6): 684-689.

Ice, G.G., and J. D. Stednick (eds.). 2004. A Century of Forest and Wildland Watershed Lessons. Bethesda, MD: Society of American Foresters.

Ice, G.G. and J.D. Stednick. 2004. A Century of Forest and Wildland Watershed Lessons. Bethesda, MD: Society of American Foresters.

Ice, G.G., P.W. Adams, R.L. Beschta, H.A. Froehlich, and G.W. Brown. 2004. Forest Management to Meet Water Quality and Fisheries Objectives: Watershed Studies and Assessment Tools in the Pacific Northwest. Pp. 239-262 In Ice, G.G., and J.D., Stednick (eds.) A Century of Forest and Wildland Watershed Lessons.. Bethesda, MD: Society of American Foresters.

Johnson, E.A. 1952. Effect of farm woodland grazing on watershed values in the southern Appalachian Mountains. Journal of Forestry 50(2):109-113.

Johnson, E.A., K. Miyanishi, and S.R.J. Bridge. 2001. Wildfire regime in the boreal forest and the udea of suppression and fuel buildup. Conservation Biology 15(6):1554–1557.

Johnson, M.G., and R.L. Beschta. 1980. Logging, infiltration capacity, and surface erodibility in western Oregon. Journal of Forestry 78(6):334-337(4).

Johnson, R. 1998. The forest cycle and low river flows: a review of UK and international studies. Forest Ecology and Management 109(1-3):1-7.

Johnson, S.L. 2004. Factors influencing stream temperatures in small streams:

substrate effects and a shading experiment. Canadian Journal of Fisheries and Aquatic Sciences 61:913-923.
Johnson, S.L., and J.A. Jones. 2000. Stream temperature responses to forest harvest and debris flows in western Cascades, Oregon. Canadian Journal of Fisheries and Aquatic Sciences 57(Suppl. 2): 30–39.
Jones, J.A. 2000. Hydrologic processes and peak discharge response to forest removal, regrowth, and roads in 10 small experimental basins, western Cascades, Oregon. Water Resources Research 36: 2621-2642.
Jones, J.A. 2005. Intersite comparisons of rainfall-runoff processes. Pp. 1839-1854 In Anderson, M.G. (ed.) Encyclopedia of Hydrological Sciences. New York: John Wiley & Sons, Ltd.
Jones, J.A., and D.A. Post. 2004. Seasonal and successional streamflow response to forest cutting and regrowth in the northwest and eastern United States. Water Resources Research 40:W05203.
Jones, J.A., and G.E. Grant. 1996. Peak flow response to clearcutting and roads in small and large basins, western Cascades, Oregon. Water Resources Research 32:959-974.
Kasischke, E.S., T.S. Rupp, and D.L. Verbyla. 2006. Fire trends in the Alaskan boreal forest. Pp. 285-301 In Chapin, F.S. III, M.W. Oswood, K, Van Cleve, L.A. Viereck, and D.L. Verbyla (eds.) Alaska's Changing Boreal Forest. New York: Oxford University Press.
Kenney, D.S., S.T. McAllister, W.H. Caile, and J.S. Peckham. 2000. The New Watershed Source Book. Boulder, CO: Natural Resources Law Center.
Keppeler, E.T., and R.R. Ziemer. 1990. Logging Effects on Streamflow: Water Yield and Summer Low Flows at Caspar Creek in Northwestern California. Water Resources Research 25(7):1669-1679.
King, J.G., and L.C. Tennyson 1984. Alteration of streamflow characteristics following road construction in north central Idaho. Water Resources Research 20:1159-1163.
King. R.S., M.E. Baker, D.F. Whigham, D.E. Weller, T.E. Jordan, P.F. Kazyak, and M.K. Hurd. 2005. Spatial considerations for linking watershed land cover to ecological indicators in streams. Ecological Applications 15:137-153.
Klock, G.O. 1975. Impact of five postfire salvage logging systems on soils and vegetation. Journal of Soil and Water Conservation 30:78-81.
Kochenderfer, J.N., P.J. Edwards, and F. Wood. 1997. Hydrologic impacts of logging an Appalachian watershed using West Virginia's Best Management Practices. Northern Journal of Applied Forestry 14(4):207-218.
Krammes, J.S., and R.M. Rice. 1963. Effects of fire on the San Dimas Experimental Forest. Pp. 31-34 In Proceeding, 7th Annual Meeting, Arizona Watershed Symposium, 18 September, Phoenix.
Kunze, M.D., and J.D. Stednick. 2006. Streamflow and suspended sediment yield following the 2000 Bobcat fire, Colorado. Hydrological Processes 20:1661-1681.

LaMarche, J.L., and D.P. Lettenmeier. 2001. Effects of forest roads on flood flows in the Deschutes River, Washington. Earth Surface Processes and Landforms 26(2):115-134.

Landres, P.B., P. Morgan, and F.J. Swanson. 1999. Overview of the use of natural variability concepts in managing ecological systems. Ecological Applications 9(4):1179-1188.

Landsberg, J.D., and A.R. Tiedemann. 2000. Fire Management. Pp 124-138 in Drinking water from forests and grasslands: A synthesis of the scientific literature, G.E. Dissmeyer, ed. Gen. Tech. Rep. SRS–39. Asheville, NC: U.S. Department of Agriculture, Forest Service, Southern Research Station.

Larsen, I.J., and L.H. MacDonald. 2007. Predicting postfire sediment yields at the hillslope scale: Testing RUSLE and Disturbed WEPP. Water Resources Research 43, W11412, doi:10.1029/2006WR005560.

Larsen, M.C., and J.E. Parks. 1998. How wide is a road? The association of roads and mass-wasting in a forested montane environment. Earth Surface Processes and Landforms 22(9):835-848.

Lawrence, G.B., and T.G. Huntington. 1999. Soil-Calcium depletion linked to acid rain and forest growth in the Eastern United States. U.S. Department of the Interior WRIR 98-4267 U.S. Geological Survey. Available online at *http://bqs.usgs.gov/acidrain.*

Lawrence, G.B., M.B. David, S.W. Bailey, and W.C. Shortle. 1997. Assessment of calcium status in soils of red spruce forests in the northeastern United States. Biogeochemistry 38:19-39.

Lee, P., C. Smyth, and S. Boutin. 2004. Quantitative review of riparian buffer width guidelines from Canada and the United States. Journal of Environmental Management 70:165-180.

Lefsky, M. A., D. J. Harding, M. Keller, W. B. Cohen, C. C. Carabajal, F. Del Bom Espirito-Santo, M. O. Hunter, and R. de Oliveira Jr. 2005. Estimates of forest canopy height and aboveground biomass using ICESat. Geophys. Res. Lett. 32, L22S02, doi:10.1029/2005GL023971.

Lefsky, M.A., W.B. Cohen, G.G. Parker and D.J. Harding. 2002. Lidar remote sensing for ecosystem studies. Bioscience 52:19-30.

Leith, R.M.M., and P.H. Whitfield. 1998. Evidence of climate change effects on the hydrology of streams in South-Central B.C. Canadian Water Resources Journal 23(3):219-230.

Leopold, L.B., M.G.Wolman, and J.P. Miller. 1964. Fluvial Processes in Geomorphology. San Francisco: W.H. Freeman and Co..

Letey, J. 2001. Causes and consequences of fire-induced soil water repellency. Hydrological Processes 15:2867-2875.

Levno, A., and J. Rothacher. 1967. Increases in maximum stream temperatures after logging in old-growth Douglas-fir watersheds. Res. Note PNW-65. Portland, OR: U.S. Department of Agriculture, Forest Service, Pacific Northwest Forest and Range Experiment Station.

Levno, A., and J. Rothacher. 1969. Increases in maximum stream temperatures after slash burning in a small experimental watershed. Res. Note PNW-110.

Portland, OR: U.S. Department of Agriculture, Forest Service, Pacific Northwest Forest and Range Experiment Station.

Lewis, G.P., and G.E. Likens. 2007. Changes in stream chemistry associated with insect defoliation in a Pennsylvania hemlock-hardwoods forest. Forest Ecology and Management 238(1-3):119-211.

Libohova, Z. 2004. Effects of thinning and a wildfire on sediment production rates, channel morphology, and water quality in the Upper South Platte River watershed. M.S. Thesis. Colorado State University, Fort Collins, CO.

Light, S. S., J. R. Wodraska, and J. Sabina. 1989. The Southern Everglades: The Evolution of Water Management. The National Forum, Winter. 8.

Likens, G.E., and F.H. Bormann. 1995. Biogeochemistry of a Forested Ecosystem. New York: Springer-Verlag.

Likens, G.E., C.T. Driscoll, and D.C. Buso, 1996. Long-term effects of acid rain: response and recovery of a forested ecosystem. Science 272: 244–246.

Likens, G.E., F.H. Bormann, N.M. Johnson, D.W. Fisher, and R.S. Pierce. 1970. Effects of forest cutting and herbicide treatment on nutrient budgets in the Hubbard Brook watershed ecosystem. Ecological Monograph 40(1): 23–47.

Likens, G.E., F.H. Bormann, R.S. Pierce, J.S. Eaton, and N.H. Johnson. 1977. Biogeochemistry of a Forested Ecosystem. New York: Springer-Verlag.

Logan, J.A., J. Regniere, and J.A. Powell. 2003. Assessing the impacts of global warming on forest pest dynamics. Frontier Ecology and Environment 1(3):130-137.

Love, L.D. 1955. The effect on streamflow of the killing of spruce and pine by the Engelmann spruce beetle. Transactions of the American Geophysical Union 36:113-118.

Lovett, G.M., K.C. Weathers, and W.V. Sobczak. 1999. Nitrogen saturation and retention in forested watersheds of the Catskill Mountains, NY. Ecological Applications 10(1):73-84.

Lovett, G.M., and J.D. Kinsman. 1990. Atmospheric pollutant deposition to high-elevation ecosystems. Atmospheric Environment 24A:239–264.

Lovett, G.M., and S. E. Lindberg. 1993. Atmospheric deposition and canopy interactions of nitrogen in forests. Canadian Journal of Forest Research 23:1603–1616.

Lowrance, R., L.S. Altier, J.D. Newbold, R.R. Schnabel, P.M. Groffman, J.M. Denver, D.L. Correll, J.W. Gilliam, J.L. Robinson, R.B. Brinsfield, K.W. Staver, W. Lucas, A.H. Todd. 1997. Water quality functions of riparian forest buffers in Chesapeake Bay watersheds. Environmental Management 21(5):687-712.

Luce, C.H., and T.W. Cundy.1994. Parameter identification for a runoff model for forest roads. Water Resources Research 30:1057–1069.

Lynch, J.A., and E.S. Corbett. 1990. Evaluation of best management practices for controlling nonpoint pollution from silvicultural operations. Journal of the American Water Resources Association 26(1):41–52.

Lynch, J.A., E.S. Corbett, and K. Mussallem. 1985. Best management practices for controlling nonpoint-source pollution on forested watersheds. Journal of Soil and Water Conservation 40(1):164-167.

MacDonald, J.S., P.G. Beaudry, E.A. MacIsaac, and H.E. Herunter. 2003. The effects of forest harvesting and best management practices on streamflow and suspended sediment concentrations during snowmelt in headwater streams in sub-boreal forests of British Columbia, Canada. Canadian Journal of Forest Research 33(8):1397–1407.

Macdonald, K.B., and F. Weinmann (eds.). 1997. Wetland and Riparian Restoration: Taking a Broader View. Contributed Papers and Selected Abstracts. Society for Ecological Restoration, 1995 International Conference. September 14-16, 1995. University of Washington. Seattle, Washington, USA. Publication EPA 910-R-97-007. Seattle, WA: USEPA, Region 10.

MacDonald, L.H., and D. Coe. 2007. Influence of headwater streams on downstream reaches in forested areas. Forest Science 53(2):148-168.

MacDonald, L.H. 2000. Evaluating and managing cumulative effects: process and constraints. Environment Management 26(3):299-315.

MacDonald, L.H., and D. Coe. 2007. Influence of headwater streams on downstream reaches in forested areas. Forest Science 53(2):148-158.

MacDonald, L.H., and E.L. Huffman. 2004. Post-fire soil water repellency: Persistence and soil moisture thresholds. Soil Science Society of America Journal 68:1729-1734.

MacDonald, L.H., and J.D. Stednick. 2003. Forests and water: A state-of-the-art review for Colorado. Colorado Water Resources Research Institute Report no. 196. Denver, CO: Colorado Water Resources Research Institute.

Mack, M.C., and C.M. D'Antonio. 1998. Impacts of biological invasions on disturbance regimes. Trends in Ecology an Evolution 13(5):195-198.

Madej, M. 2001. Erosion and sediment delivery following removal of forest roads. Earth Surf. Process. Landforms 26:175–190.

Martin, C.W., and R.D. Harr. 1989. Logging of mature Douglas-fir in western Oregon has little effect on nutrient output budgets. Canadian Journal of Forest Research 19:35–43.

Martin, C.W., J.W. Hornbeck, G.E. Likens, and D.C. Buso. 2000. Impacts of intensive harvesting on hydrology and nutrient dynamics of northern hardwood forests. Canadian Journal of Fisheries and Aquatic Sciences 57(Suppl. 2): 19–29.

May, C.L. 2002. Debris flows through different forest age classes in the central Oregon Coast Range. Journal of the American Water Resources Association 38(4):1097–1113.

McBroom, M.W., R. S. Beasley, M. Chang, and G.G. Ice. 2008. Water quality effects of clearcut harvesting and forest fertilization with Best Management Practices. Journal of Environmental Quality 37:114-124.

McCoy, J., P. Pannill, and S. Barker. 2000. Evaluating the Effectiveness of Maryland's Best Management Practices for Forest Harvesting Operations. Maryland Dept. Natural Resources Report FWHS-FS-00-01. Annapolis,

MD. Available online at http://www.dnr.state.md.us/forests/mbmp/. Last accessed May 5, 2008.

McCullough, D.G., R.A. Wener, and D. Neumann. 1998. Fire and insects in northern and boreal forest ecosystems of North America. Annual Review of Entomology 43: 107-127.

McIver, J.D., and R. McNeil. 2006. Soil disturbance and hill-slope sediment transport after logging of a severely burned site in northeastern Oregon. Western Journal of Applied Forestry 21:123-133.

Megahan, W.F. 1972. Subsurface flow interception by a logging road in mountains of central Idaho. National Symposium on Watersheds in Transition Colorado State University, American Water Resources Association.

Megahan, W.F. 1983. Hydrologic effects of clearcutting and wildfire on steep granitic slopes in Idaho. Water Resources Research 19:811-819.

Megahan, W.F., and J. Hornbeck. 2000. Lessons Learned in Watershed Management: A Retrospective View. USDA Forest Service Proceedings RMRS–P–13 177-188.

Megahan, W.F., M. Wilson, S.B. Monsen 2001. Sediment production from granitic cutslopes on forest roads in Idaho, USA. Earth Surface Processes and Landforms 26: 153-163.

Megahan, W.F., N.F. Day, and T.M. Bliss. 1978. Landslide occurrence in the western and central Rocky Mountain physiographic province in Idaho. Pp. 115-139 In Proceedings of the 5th North American Soils Conference, Colorado State University, Fort Collins, Colorado.

Melillo, J.M., A.S. McGuire, D.W. Kickllighter, B. Moore, C.J. Vorosmarty, and A.L. Schloss. 1993. Global climate change and terrestrial net primary production. Nature 363:234-240.

Meyer, J.L., and G.E. Likens. 1979. Transport and transformation of phosphorus in a forest stream ecosystem. Ecology 60: 1255-1269.

Michael, J.L. 2000. Pesticides. Pp 139-152 in Drinking water from forests and grasslands: a synthesis of the scientific literature. In Dissmeyer, G.E. (ed) Gen. Tech. Rep. SRS–39. Asheville, NC: U.S. Department of Agriculture, Forest Service, Southern Research Station.

Millar, C.I., M.L. Stephenson, and S.L. Stephens. 2008. Climate change and forests of the future: managing in the face of uncertainty. Ecological Applications 17(8): 2145–2151.

Miller, D.J., and K.M. Burnett, 2007. Effects of forest cover, topography, and sampling extent on the measured density of shallow, translational landslides. Water Resources Research 43:Wo433.

Miller, J.H., and M. Newton. 1983. Nutrient losses from disturbed forest watersheds in Oregon's coast range. Agro-Ecosystems 8:153–167.

Miller, N.L., K.E. Bashford, and E. Strem. 2003. Potential impacts of climate change on California hydrology. Journal of the American Water Resources Association 39(4):771–784.

Milly, P.C.D. J. Betancourt, M. Falkenmark, R.M. Hirsch, Z.W. Kundzewicz, D.P. Lettenmaier, and R.J. Stouffer. 2008. Climate change - Stationarity is dead: Whither water management? Science 319:573-574.

Montgomery, D.R., K.M. Schmidt, H.M. Greenberg, and W.E. Dietrich, 2000. Forest clearing and regional landsliding. Geology 28(4):311-314.

Moody, J.A., and D.A. Martin. 2001. Initial hydrologic and geomorphic response following a wildfire in the Colorado Front Range. Earth Surface Processes and Landforms 26:1049-1070.

Moody, J.A., J.D. Smith, and B.W. Ragan. 2005. Critical shear stress for erosion of cohesive soils subjected to temperatures typical of wildfires. Journal of Geophysical Research 110:F01001.

Moore, G.W., B.J. Bond, J.A. Jones, N.Phillips, and F.C. Meinzer. 2004. Structural and compositional controls on transpiration in a 40- and 450-yr-old riparian forest in western Oregon, USA. Tree Physiology 24:481-491.

Moore, R.D., and S.M. Wondzell. 2005. Physical hydrology and the effects of forest harvesting in the Pacific Northwest: a review. Journal of the American Water Resources Association 41(4):763-784.

Morgenstern, K. 2006. Summary of Issues and Concepts Hydrologic Impacts of Forest Management on Municipal Water Supplies and Hydro-Electric Generation. Presented to NRC Committee on Forest Hydrology, Eugene, OR, September 27, 2006.

Mortimer, M.J., and R.J.M. Visser. 2004. Timber harvesting and flooding: emerging legal risks and potential mitigations. South. J. Applied Forest 28:69-75.

Moser, B., M. Scutz M, K.E. Hindenlang. 2008. Resource selection by roe deer: Are windthrow gaps attractive feeding places? Forest Ecology And Management 255:1179-1185.

Naiman, R.J., and H. Decamps. 1997. The ecology of interfaces: Riparian zones. Annual Review of Ecology and Systematics 28:621-658.

NRC (National Research Council). 1999. New Strategies for America's Watersheds. Washington, DC: National Academy Press.

NRC. 2000. Watershed Management for Potable Water Supply. Washington, DC: National Academy Press

NRC. 2002. The Missouri River Ecosystem: Exploring the Prospects of Recovery. Washington, DC: National Academies Press.

National Council for Air and Stream Improvement (NCASI). 1999. Water quality effects of forest fertilization. NCASI Tech. Bull. 782. New York: National Council for Air and Stream Improvement.

Neary, D.G., C.C. Klopatek, L.F. DeBano, and P.F. Ffolliott. 1999. Fire effects on belowground sustainability: A review and synthesis. Forest Ecology and Management 122:51-71.

Neary, D.G., and P.F. Ffolliott. 2005. The water resource: Its importance, characteristics, and general response to fire. Pp. 95-106 In Part B. In Neary, D.G., K.C. Ryan, and L.F. DeBano (eds.) Wildland Fire in Ecosystems: Effects of Fire on Soil and Water. General Technical Report RMRS-GTR-42-

vol. 4. Fort Collins, CO: US Forest Service, Rocky Mountain Research Station.
Neary, D.G., J.D. Landsberg, A.R. Tiedemann, and P.F. Ffolliott. 2005c. Water quality. Pp. 119-134 In Neary, D.G., K.C. Ryan, and L.F. DeBano(eds) Wildland Fire in Ecosystems: Effects of Fire on Soil and Water. USDA Forest Service General Technical Report RMRS-GTR-42. Volume 4.
Neary, D.G., K.C. Ryan, and L.F. DeBano (eds.). 2005b. Wildland Fire in Ecosystems: Effects of Fire on Soil and Water. USDA Forest Service General Technical Report RMRS-GTR-42, Volume 4.
Neary, D.G., P.F. Ffolliott, and J.D. Landsberg. 2005b. Fire and streamflow regimes. Pp. 107-118 In Neary, D.G., K.C. Ryan, and L.F. DeBano (eds.) 2005. Wildland Fire in Ecosystems: Effects of Fire on Soil and Water. USDA Forest Service General Technical Report RMRS-GTR-42, Volume 4.
Neitsch, S.L., J.G. Arnold, J.R. Kiniry, J.R. Williams and K.W. King. 2002. Soil and water assessment tool – theoretical documentation. Version 2000. Grassland, Soil and Water Research Laboratory. Temple, TX: Agricultural Research Service.
Northwest Plan (U.S. Dept. of Agriculture). 1996. The Northwest Forest Plan: A Report to the President and Congress. Office of Forestry and Economic Assistance. E. Thomas Tuchmann [et al.] Portland, OR: The Office, [1996] xi, 253 p. Appendix A: The Forest Plan for a Sustainable Economy and a Sustainable Environment to President William J. Clinton, Vice President Albert Gore, Jr.
Parendes, L.A., and J.A. Jones. 2000. Role of light availability and dispersal in exotic plant invasion along roads and streams in the H. J. Andrews Experimental Forest, Oregon Conservation Biology 14:64-75.
Parmesan, C., and G. Yohe. 2003. Globally coherent fingerprint of climate change impacts across natural systems. Nature 421:37-42.
Payne, J.T., A.W. Wood, A.F. Hamlet, R.N. Palmer, and D.P. Lettenmaier. 2004. Mitigating the effects of elimate ehange on the water resources of the Columbia River basin. Climatic Change 62(1-3):233-256.
Pearson, R.G., and T.P. Dawson. 2003. Predicting the impacts of climate change on the distribution of species: Are bioclimate envelope models useful? Global Ecology and Biogeography 12(5):361–371.
Perera, A. H., L. J. Buse, and M G. Weber (eds.). 2004. Emulating Natural Forest Landscape Disturbances: concepts and applications. New York, NY: Columbia University Press.
Perry, T.D. 2007. Do vigorous young forests reduce streamflow? Results from up to 54 years of streamflow records in eight paired-watershed experiments in the H. J. Andrews and South Umpqua Experimental Forests. MS thesis, Oregon State University.
Peterjohn, W.T., and D.L. Correll. 1984. Nutrient dynamics in an agricultural watershed: Observations on the role of a riparian forest. Ecology 65(5):1466-1475.

Platts, W.S. 1981. Influence of Forest and Rangeland Management on Anadromous Fish Habitat in Western North America: Effects of livestock grazing. USDA Forest Service Pacific Northwest Forest and Range Experiment Station General Technical Report PN W-124.

Poff, N.L., J.D. Allan, M.B. Bain, J.R. Karr, K.L. Prestegaard, B.D. Richter, R.E. Sparks, and J.C. Stromberg. 1997. The natural flow regime. Bioscience 47: 769-784.

Polis, G.A., A.L.W. Sears, G.R. Huxel, D.R. Strong, and J. Maron. 2000. When is a trophic cascade a trophic cascade? Trends in Ecology and Evolution 15:473-475.

Powell, D.S., J.L. Faulkner, D.R. Darr, Z. Zhu, and D.W. MacCleery. 1993. Forest Resources of the United States, 1992. Gen. Tech. Rep. RM-234. Fort Collins, CO: U.S. Department of Agriculture, Rocky Mountain Forest and Range Experiment Station.

Powers, J.S., P. Sollins, M.E. Harmon, and J.A. Jones. 1999. Plant-pest interactions in time and space: A Douglas-fir bark beetle outbreak as a case study. Landscape Ecology 14:105-120.

Prentice, I.C., M.T. Sykes, and W. Cramer. 1993. A simulation model for the transient effects of climate change on forest landscapes. Ecological Modelling 65(1-2):51-70.

Quabbin, L.M.P. 2007. Quabbin Reservoir Watershed System: Land Management Plan 2007-2017. Section 5: Management Plan Objectives and Methods.

Radeloff, V.C., R.B. Hammer, S.I. Stewart, J.S. Fried, S.S. Holcomb, and J.F. McKeefry. 2005. The wildland-urban interface in the US. Ecological Applications 15(3):799-805.

Ranalli, A.J. 2004. A Summary of the Scientific Literature on the Effects of Fire on the Concentration of Nutrients in Surface Water, U.S. Geological Survey Open-File Report 2004-1296. Available online at http://pubs.water.usgs.gov/ofr2004-1296.

Reeves, G.H., J.E. Williams, K.M. Burnett, and K. Gallo. 2006. The aquatic conservation strategy of the Northwest Forest Plan. Conservation Biology 20:319-329.

Reid, L.M. 1993. Research and cumulative watershed effects. Gen. Tech. Rep. PSW-GTR-141. Albany, CA: Pacific Southwest Research Station, Forest Service, U.S. Department of Agriculture.

Reid, L.M., and J. Lewis, 2007. Rates and implication of rainfall interception in a coastal redwood forest. USDA Forest Service PSW-GTR 194:107-117.

Reid, L.M., and T. Dunne. 1984. Sediment production from forest road surfaces. Water Resources Research 20(11):1753-1761.

Reid, L.M., and T. Dunne. 1996. Rapid Construction of Sediment Budgets for Drainage Basins. Cremlingen, Germany: Catena-Verlag.

Reiter, M. 2008. December 1-4, 2007 storm events summary. Unpublished report prepared for Weyerhaeuser Western Timberlands. Feb 8.

Renard, K. G., G. R. Foster, G. A. Weeises, D. K. McCool, and D. C. Yoder, coordinators. 1997. Predicting soil erosion by water: a guide to conservation planning with the Revised Universal Soil Loss Equation (RUSLE). U.S. Department of Agriculture, Agriculture Handbook No. 703.

Rice, T. and J. Souder. 1998. Pulp friction and the management of Oregon's state forests. Journal of Environmental Law and Litigation 13:209-273.

Riggan, P.J., R.N. Lockwood, and E.N. Lopez. 1985. Deposition and processing of airborne nitrogen pollutants in Mediterranean-type ecosystems of southern California. Environmental Science and Technology 19:781–789.

Rinne, J.N., and W.L. Minckley. 1991. Native fishes of arid lands: A dwindling resources of the Desert Southwest. USDA Forest Service, General Technical Report RM-206.

Rinne, J.N. 1996. Sort-term effects of wildfire on fishes and aquatic macrovertebrates: Southwestern United States. Journal of Fisheries Management 16:653-658.

Ripple, W.J., and R.L. Beschta. 2007. Restoring Yellowstone's aspen with wolves. Biological Consesrvation 138:514-519.

Risser, P.G. 1995. The status of the science examining ecotones. BioScience 45(5):318-325.

Robichaud, P.R. 2000. Fire effects on infiltration rates after prescribed fire in Northern Rocky Mountain forests, USA. Journal of Hydrology 231-232:220-229.

Robichaud, P.R., and R.E. Brown. 1999. What happened after the smoke cleared: onsite erosion rates after a wildfire in eastern Oregon. Pp. 419-426 In Olsen, D., and J.P. Potyondy (eds.) Proceedings of the American Water Resources Association Specialty Conference on Wildland Hydrology. Herdon, VA: American Water Resources Association.

Rummell, R.S. 1951. Some Effects of Livestock Grazing on Ponderosa Pine Forest and Range in Central Washington. Ecology 32(4):594-607.

Running, S.W., and R.R. Nemani. 1991. Regional hydrologic and carbon balance responses of forests resulting from potential climate change. Climatic Change 19(4):349-368.

Rustad, L.E., I.J. Fernandez, M.B. David, M.J. Mitchell, K.J. Nadelhoffer, and R.B. Fuller. 1996. Experimental soil acidification and recovery at the Bear Brook watershed in Maine. Soil Science Society of America Journal 60:1933-1943.

Ryan, D.F., and S. Glasser. 2000. Goals of this Report. Pp 3-6 in Drinking water from forests and grasslands: a synthesis of the scientific literature, G.E. Dissmeyer, ed. Gen. Tech. Rep. SRS–39. Asheville, NC: U.S. Department of Agriculture, Forest Service, Southern Research Station.

Sabatier, P., W. Focht, M. Lubell, Z. Trachterberg, A. Vedlitz, and M. Matlock. 2005. Swimming upstream: collaborative approaches to watershed management. Cambridge, MA: MIT Press.

Sahin, V., and M. J. Hall. 1996. The effects of afforestation and deforestation on water yields. Journal of Hydrology 178 (1/4):293–309.

Sass, G.Z., and I.F. Creed. 2007. Characterizing hydrodynamics on boreal landscapes using archived synthetic aperture radar imagery. Hydrological Processes.

Sass, G.Z., and I.F. Creed. 2008. Characterizing hydrodynamics on boreal landscapes using archived synthetic aperture radar imagery. Hydrological Processes 22: 1687-1699.

Satterlund, D. R., and P. W. Adams. 2001. Wildland Watershed Management. New York: John Wiley & Sons.

Scatena, F.N. 2000. Drinking Water Quality. Pp 7-25 In Dissmeyer, G.E. (ed) Drinking Water from Forests and Grasslands: A Synthesis of the Scientific Literature. General Technical Report SRS–39. Asheville, NC: U.S. Department of Agriculture, Forest Service, Southern Research Station

Schmid, J.M., and S.A. Mata. 1996. Natural Variability of Specific Forest Insect Populations and Their Associated Effects in Colorado. USDA Forest Service General Technical Report RM-GTR-275.

Schmid, J.M., S.A. Mata, M.H. Martinez, and C.A. Troendle. 1991. Net Precipitation Within Small Group Infestations of the Mountain Pine Beetle. USDA Forest Service Research Note RM-508. Fort Collins, CO: Rocky Mountain Forest and Range Experiment Station.

Schnorbus, M., and Y. Alila. 2004. Forest Harvesting Impacts on the Peak Flow Regime in the Columbia Mountains of Southeastern British Columbia: An Investigation Using Long-Term Numerical Modeling. Water Resources Research 40:W05205.

Schuler, J.L., and R.D. Briggs. 2000. Assessing Application and Effectiveness of Forestry Best Management Practices in New York. Northern Journal of Applied Forestry 17(4):125-134(10).

Schullery, P. 1986. The Bears of Yellowstone. Boulder, CO: Roberts Rinehart Inc.

Scott, D.F. 1993. The hydrological effects of fire in South African mountain catchments. Journal of Hydrology 150:409-432.

Seattle Times. December 16, 2007. Local news, page B-1.

Seyedbagheri, K.A. 1998. Idaho forestry best management practices: compilation of research on their effectiveness. General Technical Report INT-GTR-339. Ogden, UT: USDA Forest Service Intermountain Station Region. 66-67.

Shafroth, P.B., J.R. Cleverly, T.L. Dudley, J.P. Taylor, C. Van Riper III, E.P. Weeks, and J.N. Stuart. 2005. Control of Tamarix in the western US: Implications for water salvage, wildlife Use, and riparian restoration. Environmental Management 35(3):231 246.

Shakesby, R.A., and Doerr, S.H. 2006. Wildfire as a hydrological and geomorphological agent. Earth-Science Reviews 74:269-307.

Shepherd, K.D. and M.G. Walsh. 2002. Development of reflectance spectral libraries for characterizing soil properties. Soil Science Society of America Journal 66(3):988-998.

Shepherd, K.D. and M.G. Walsh. 2004. Diffuse reflectance spectroscopy fro rapid soil analysis. In Encyclopedia of Soil Science. New York: Marcel Dekker Inc.

Sidle, R. C. 2000. Watershed challenges for the 21st century: A global perspective for mountainous terrain. Pp. 46-56 In Ffolliott, P. F., M. B. Baker, Jr., C. B. Edminster, M. C. Dillon, and K. L. Mora (technical coordinators) Land stewardship in the 21st century: The contributions of watershed management. USDA Forest Service, Proceedings RMRS-P-13.

Sidle, R.C. 2006. Field observations and process understanding in hydrology: essential components in scaling. Hydrological Processes 20:1439-1445.

Sidle, R.C., A.J. Pearce, and C.L. O'Loughlin. 1985. Soil mass movement—Influence of natural factors and land use. Water Resource Monograph 11. Washington, DC: American Geophysical Union.

Sidle, R.C., and H. Ochiai, 2006. Landslides: Processes, prediction, and land use. Water Resources Monograph 18. Washington, DC: American Geophysical Union.

Sinclair, J.D., and E.L. Hamilton. 1955. Streamflow reactions in a fire-damaged watershed. Proceedings American Society of Civil Engineers, Hydraulics Division 81(629):1-17.

Singh, R.K., V.H. Prasad, and C.M. Bhatt. 2004. Remote sensing and GIS approach for assessment of the water balance of a watershed. Hydrological Sciences Journal 49:131-141.

Singh, V.P. (ed.) 1995. Computer Models of Watershed Hydrology. Highlands Ranch, CO: Water Resources Publications.

Singh, V.P., and D.K. Frevert (eds.) 2006. Watershed Models. Boca Raton: CRC Press.

Sinton, D.S., J.A. Jones, J.L. Ohmann, and F.J. Swanson. 2000. Windthrow disturbance, forest composition, and structure in the Bull Run Basin. Oregon. Ecology 81(9):2539-2556.

Sivapalan, M. 2003. Prediction in ungauged basins: a grand challenge for theoretical hydrology. Hydrological Processes 17:3163-3170.

Smith, W.B., P.D. Miles, J.S. Vissage, and S.A. Pugh. 2004. Forest resources of the United States, 2002. USDA Forest Service, General Technical Report NC-241.

Spencer, C.N., K. Odney- Gabel, and F.R. Hauer. 2003. Wildfire effects on stream food webs and nutrient dynamics in Glacier National Park, USA. Forest Ecology and Management 178: 141-153.

Stednick , J.D., C.A. Troendle and G.G. Ice. 2004. Lessons for watershed research in the future. Pp. 277-287 In Ice, G.G. and J.D. Stednick (eds.) A Century of Forest and Wildland Watershed Lessons. Bethesda, MD: Society of American Foresters.

Stednick, J.D. 1996. Monitoring the effects of timber harvest on annual water yield. Journal of Hydrology 176 (1/4):79–95.

Stednick, J.D. 2000. Timber Management. Pp 103-119 In Dissmeyer, G.E. (ed) Drinking Water from Forests and Grasslands: A Synthesis of the Scientific

Literature Gen. Tech. Rep. SRS–39. Asheville, NC: U.S. Department of Agriculture, Forest Service, Southern Research Station.

Stocks, B.J., M.A. Fosberg, T.J. Lynham, L. Mearns, B.M. Wotton, Q. Yang, J-Z. Jin, K. Lawrence, G. R.Hartley, J. A. Mason, and D.W. McKenney. 1998. Climate change and forest fire potential in Russian and Canadian boreal forests. Climatic Change 38(1):1-13.

Story, A., R.D. Moore, and J.S. Macdonald. 2003. Stream temperatures in two shaded reaches below cutblocks and logging roads: downstream cooling linked to subsurface hydrology. Canadian Journal of Forest Research 33(8):1383–1396.

Swank, W. 2000. Forest Succession. Pp 120-123 In Dissmeyer, G.E. (ed) Drinking Water from Forests and Grasslands: A Synthesis of the Scientific Literature. Gen. Tech. Rep. SRS–39. Asheville, NC: U.S. Department of Agriculture, Forest Service, Southern Research Station.

Swank, W.T. 1988. Stream chemistry responses to disturbance. Pp. 339-357 In Swank, W. and D.A. Crossley, Jr. (eds.) Forest Hydrology and Ecology at Coweeta. Ecological Studies Volume 66. New York: Springer-Verlag.

Swank, W.T., and C.A. Crossley Jr.(eds). 1988. Forest Hydrology and Ecology at Coweeta. New York: Springer-Verlag.

Swank, W.T., and J.M. Vose. 1997. Long-term nitrogen dynamics of Coweeta forested watersheds in the Southeastern United States of America. Global Biogeochemical Cycles 11(4): 657–671.

Swank, W.T., J.M. Vose, and K.J. Elliott. 2001. Long-term hydrologic and water quality responses following commercial clearcutting of mixed hardwoods on a southern Appalachian catchment. Forest Ecology and Management 143(1-3):163-178.

Swank, W.T., L.W. Swift, Jr., and J.E. Douglass. 1988. Streamflow changes associated with forest cutting, species conversions, and natural disturbances. Pp. 297-312 In Swank, W., and D.A. Crossley, Jr.(eds) Forest Hydrology and Ecology at Coweeta. New York: Springer-Verlag.

Swanson, F.J., and J.F. Franklin. 1992. New Forestry Principles from Ecosystem Analysis of Pacific Northwest Forests. Ecological Applications 2(3):262-274.

Swanson, F.J., and C.T. Dyrness. 1975. Impact of clear-cutting and road construction on soil erosion by landslides in the western Cascade Range, Oregon. Geology3(7):393-396.

Swanson, F.J., F.N. Scatena, G.E. Dissmeyer, M.E. Fenn, E.S. Verry, and J.A. Lynch 2000. Watershed Processes—Fluxes of Water, Dissolved Constituents, and Sediment. Pp 26-41 In Dissmeyer, G.E. (ed.) Drinking Water from Forests and Grasslands: A Synthesis of the Scientific Literature. General Technical Report SRS–39. Asheville, NC: U.S. Department of Agriculture, Forest Service, Southern Research Station.

Swetnam, T.W. 2005. Fire histories from pine-dominated forest in the Madrean Archipelago. Pp. 35-43 In Gottfried, G. J., B. S. Gebow, L. G. Eskew, and C. B. Edminster (compilers) Connecting mountain islands and desert seas:

Biodiversity and management of the Madrean Archipelago. USDA Forest Service, Proceedings RMRS-P-36.

Swetnam, T.W., and C.H. Baison. 1996. Fire histories of montane forests in the Madrean Borderlands. Pp. 15-36 In Ffolliott, P. F., L. F. DeBano, M. B. Baker, Jr., G. J. Gottfried, G. Solis-Garza, C. B. Edminster, D. G. Neary, L. S. Allen, and R. H. Hamre (technical coordinators) Effects of Fire in the Madrean Province Ecosystems: A Symposium Proceedings. USDA Forest Service, General Technical Report RM-GTR-289

Switalski, T.A., J.A. Bissonette, T.H. DeLuca, C.H. Luce, and M.A. Madej. 2004. Benefits and impacts of road removal. Frontiers in Ecology and the Environment 2(1):21–28.

Szewczyk, R., J. Polastre, A. Mainwaring, D. Culler, 2004, Lessons learned from a sensor network expedition. Pp. 307-322 In Karl, H., A. Willig, A. Wolisz (eds.) Proceedings of Wireless Sensor Networks, First European Workshop, EWSN 2004, Berlin, Germany.

Tague, C., and L. Band. 2001. Simulating the impact of road construction and forest harvesting on hydrologic response. Earth Surface Processes and Landforms 26(2):135-152.

Tatum, V. 2003. Introduction and triclopyrtechnical. Volume I: The toxicity of sivilcultural herbicides to wildlife. Bulletin No. 861. Gainesville, FL: National Council for Air and Stream Improvement, Southern Regional Center.

Tatum, V. 2004. Glyphosate and imazapyr. Volume II: The toxicity of sivilcultural herbicides to wildlife. Bulletin No. 861. Gainesville, FL: National Council for Air and Stream Improvement, Southern Regional Center.

Taylor, P.L. 2005. In the Market But Not of It: Fair Trade Coffee and Forest Stewardship Council Certification as Market-Based Social Change. World Development 33(1):129-147.

Taylor, S.W., and A.L. Carroll. 2004. Disturbance, Forest Age, and Mountain Pine Beetle Outbreak Dynamics in BC: A Historical Perspective. Mountain Pine Beetle Symposium: Challenges and Solutions. October 30-31, 2003, Kelowna, British Columbia. T.L. Shore, J.E. Brooks, and J.E. Stone (eds.) Natural Resources Canada, Canadian Forest Service, Pacific Forestry Centre, Information Report BC-X-399, Victoria, BC.

Terborgh, J., L. Lopez, P. Nuñez, M. Rao, G. Shahabuddin, G. Orihuela, M. Riveros, R. Ascanio, G.H. Adler, T.D. Lambert, and L. Balbas. 2001. Ecological meltdown in predator-free forest fragments. Science 294 (5548):1923-1926.

Theobald, D. 2005. Landscape patterns of exurban growth in the USA from 1980 to 2020. Ecology and Society 10(1): 32.

Thomas, C.D., A. Cameron, R.E. Green, M. Bakkenes, L.J. Beaumont, Y.C. Collingham, B.F.N. Erasmus, M. Ferreira de Siqueira, A. Grainger, L. Hannah, L. Hughes, B. Huntley, A. S. van Jaarsveld, G.F. Midgley, L. Miles, M.A. Ortega-Huerta, A. Townsend Peterson, O.L. Phillips, and S.E. Williams. 2004. Extinction risk from climate change. Nature 427:145-149.

Thomas, J.W. 1996. Forest Service perspective on ecosystem management. Ecological Applications 6(3):703-705.

Thompson, J.R., T.A. Spies, and L.M. Ganio. 2007. Reburn severity in managed and unmanaged vegetation in a large wildfire. Proceedings of The National Academy of Sciences of The United States of America 104: 10743-10748.

Tiedemann, A.R. 1973. Stream chemistry following a forest fire and urea fertilization in north-central Washington. Portland, OR: USDA Forest Service, Pacific Northwest Forest and Range Experiment Station, Research Note PNW-203.

Tiedemann, A.R., C.E. Conrad, J.H. Dieterich, J.W. Hornbeck, W.F. Megahan, L.A. Viereck, and D.D. Wade. 1979. Effects of Fire on Water: a state-of-knowledge review. General Technical Report WO-10. Washington DC: USDA Forest Service.

Töyrä J, A. Pietroniro, and L.W. Martz, 2001. Multisensor hydrologic assessment of a 1797 freshwater wetland, Remote Sensing of Environment 75: 162-173.

Travis, W.R., D.M. Theobald, G.W. Mixon, and T.W. Dickinson. 2005. Western futures: A look into the patterns of land use and future development in the American West. Boulder, CO: Center of the American West, University of Colorado.

Troendle, C.A., and J.O. Reuss. 1997. Effect of clear cutting on snow accumulation and water outflow at Fraser, Colorado. Hydrology and Earth System Sciences 1(2): 325-332.

Troendle, C.A., and R.M. King. 1985. The effect of partial and clearcutting on streamflow at Deadhorse Creek, Colorado. Journal of Hydrology 90:145-157.

Troendle, C.A., M.S. Wilcox, G.S. Bevenger, and L.S. Porth, 2001. The Coon Creek water yield augmentation project: implementation of timber harvesting technology to increase streamflow. Forest Ecology and Management 143:179-187.

USDA and USDI (United States Department of Agriculture USFS and Bureau of Land Management). 2004. Record of decision. Amending resource management plans for seven Bureau of Land Management Districts and Land and Resource Management Plans for nineteen National Forests within the range of the northern spotted owl: Decision to clarify provisions relating to the Aquatic Conservation Strategy to clarify provision relating to the Aquatic Conservation Strategy. USDA USFS, Portland, Oregon and BLM, Moscow, Idaho.

USDA and USDI (United States Department of Agriculture USFS and Bureau of Land Management). 1994. Record of decision for the amendments to USFS and Bureau of Land Management Planning documents within the range of the northern spotted owl. Washington, DC: USDA USFS and BLM.

USGCRP (United States Global Change Research Program). 2000. Climate Change Impacts on the United States: The Potential Consequences of Cli-

mate Variability and Change. Overview: Forests. National Assessment Synthesis Team, US Global Change Research Program.
USGS (U.S. Geological Survey). 2004. Landslide types and processes. Fact Sheet 3072. Reston, VA: USGS.
Uunila, L., B. Guy, and R. Pike. 2006. Hydrologic effects of mountain pine beetles in the interior pine forests of British Columbia: key questions and current knowledge. Streamline 9(2):1-6.
Vagen, Tor-G, K. D. Shepherd and M.G. Walsh. 2005. Sensing landscape level change in soil fertility following deforestation and conversion in the highlands of Madagascar using Vis-NIR spectroscopy. Geoderma. Available online at www.sciencedirect.com.
Van Lear, D.H., and S.J. Danielovich. 1988. Soil movement after broadcast burning in the southern Appalachians. Southern Journal of Applied Forestry 12:49-53.
Van Lear, D.H., J.E. Douglass, S.K. Fox, and M.K. Ausberger. 1985. Sediment and nutrient export from burned and harvested pine watersheds in the South Carolina piedmont. Journal of Environmental Quality 14:169-174.
Van Sickle, J., J. Baker, A. Herlihy, P. Bayley, S. Gregory, P. Haggerty, L. Ashkenas, and J. Li. 2003. Projecting the biological condition of streams under alternative scenarios of human land use. Ecological Applications 14(2):368–380.
Verry, E.S. 1976. Estimating water yield differences between hardwood and pine forests – an application of net precipitation data. Research Paper NC-128. St. Paul, MN: USDA Forest Service, North Central Forest Experiment Station.
Verry, E.S. 1986. Forest harvesting and the Lake States experience. Water Resources Bulletin 22(6):1039-1047.
Verry, E.S., J.S. Lewis, and K.N. Brooks. 1983. Aspen clearcutting increases snowmelt and stormflow peaks in North Central Minnesota. Water Resources Bulletin 19(1):59-67.
Verry, E.S., J.W. Hornbeck, and A.C. Dolloff, editors. 2000. Riparian management in forests of the Continental Eastern United States. Boca Raton, FL: Lewis Publishers.
Vince, S.W., M.L. Duryea, E.A. Macie, and L.A. Hermansen. 2005. Forests at the wildland-urban interface: Conservation and management. Boca Raton, FL: CRC Press.
Vitousek, P. M. 1990. Biological invasions and ecosystem processes: towards an integration of population biology and ecosystem studies. Oikos 57(1):7-13.
Vitousek, P.M. 1994. Beyond global warming: ecology and global change. Ecology 75: 1861–1876.
Vowell, J.L. 2001. Using stream bioassessment to monitor best management practice effectiveness. Forest Ecology and Management 143,1-3, 237-244.

Wagenbrenner, J.W., L.H. MacDonald, and D. Rough. 2006. Effectiveness of three post-fire rehabilitation treatments in the Colorado Front Range. Hydrologic Processes 20:2989-3006.

Wagner, M. R., W. M. Block, B. W. Geils, and K. F. Wenger. 2000. Restoration ecology: A new forest management paradigm or another merit badge for foresters? Journal of Forestry 98(1):22-27.

Walter, R.C. and D.J. Merritts. 2008. Natural streams and the legacy of water-powered mills. Science 18 January 2008 319: 299-304

Walther, G.R. E. Post, P. Convey, A. Menzel, C. Parmesan, T.J.C. Beebee, J.M. Fromentin, O. Hoegh-Guldberg, and F. Bairlein. 2002. Ecological responses to recent climate change. Nature 416:389-395.

Waring, R.H., and W.H. Schlesinger. 1985. Forest Ecosystems: Concepts and Management. London: Academic Press.

Washburn, M.P. and K.J. Miller. 2003. Forest stewardship council certification. Journal of Forestry 101(8):8-13(6).

Watterson, N.A., and J.A. Jones. 2006. Flood and debris flow interactions with roads promote the invasion of exotic plants along steep mountain streams, western Oregon. Geomorphology 78:107-123.

Webster, C. R., M. A. Jenkins, and S. Jose. 2006. Woody invaders and the challenges they pose to forest ecosystems in the eastern United States. Journal of Forestry 104:366-374.

Wells, C.G., R.E. Campbell, L.F. DeBano, C.E. Lewis, R.L. Fredriksen, E.C. Franklin, R.C. Froelich, and P.H. Dunn. 1979. Effects of fire on soil, Gen. Tech. Rep. WO-7. Washington, DC: U.S. Department of Agriculture.

Wemple, B.C., and J.A. Jones. 2003. Runoff production on forest roads in a steep, mountain catchment. Water Resources Research 39(8):1220.

Wemple, B.C., F.J. Swanson, and J.A. Jones. 2001. Forest roads and geomorphic process interactions, Cascade Range, Oregon. Earth Surface Processes and Landforms 26:191-204.

Wemple, B.C., J.A. Jones, G.E. Grant 1996. Channel network extension by logging roads in two basins, western Cascades, Oregon. Water Resources Bulletin 32: 1195-1207.

Westerling, A. L., H. G. Hidalgo, D. R. Cayan, T. W. Swetnam. 2006. Warming and Earlier Spring Increases Western U.S. Forest Wildfire Activity. Science Express 313(5789): 940-943.

White, S. M. 2004. Bridging te worlds of fire managers and researchers: Lessons and opportunities from the wildland fire workshops. USDA Forest Service. General Technical Report PNW-GTR-599.

Wilcove, D.S., D. Rothstein, J. Dubow, 1998. Quantifying threats to imperiled species in the United States. Bioscience 48: 607-615.

Wilkinson, C.F., and H.M.Anderson.1985. Land and resource planning in the National Forests. Oregon Law Review 64(1-2):7-373.

Williams, J.R. 1995. The EPIC Model. Pp. 909-1000 in: Computer Models of Watershed Hydrology. Water Resources Publications: Highlands Ranch, Colorado.

Williams, M. 1989. Americans and Their Forests: A Historical Geography. New York: Cambridge University Press.

Wondzell,, S.M., and J.G. King. 2003. Postfire erosional processes in the Pacific Northwest and Rocky Mountain regions. Forest Ecology and Management 178:75-87.

Woods, S.W., A. Birkas, and R. Ahl. 2007. Spatial variability of soil hydrophobicity after wildfires in Montana and Colorado. Geomorphology 86(3-4):465-479.

Woods, S.W., and V. Balfour. 2007. An experimental study of litter and duff consumption and ash formation on post-fire runoff. Abstract H43F-1694, AGU Fall Meeting, San Francisco, CA.

Woods, S.W., R. Ahl, J. Sappington, and W. McCaughey. 2006. Snow accumulation in thinned lodgepole pine stands, Montana, USA. Forest Ecology and Management 235: 202-211.

Wright, H.A., F.M. Churchill, and W.C. Stevens. 1982. Soil loss and runoff on seeded vs. non-seeded watersheds following prescribed burning. Journal of range Management 29:294-298.

Wright, K.A., K.H. Sendek, R.M. Rice, and R.B. Thomas. 1990. Logging Effects on Streamflow: Storm Runoff at Caspar Creek in Northwestern California. Water Resources Research 26(7):1657-1667

WWPRAC (Western Water Policy Review Advisory Committee). 1998. Water in the West: Challenges for the next century. Springfield, VA: National Technical Information Service.

Wynn, T.M., S. Mostaghimi, J. W. Frazee, P. W. McClellan, R. M. Shaffer, and W. M. Aust. 2000. Effects of harvesting best management practices on surface water quality in the Virginia coastl plain. Transactions of the ASAE 43(4): 927-936.

Young, J. A., and C. D. Clements. 2005. Exotic and invasive herbaceous range plants. Rangelands 27(5):10-16.

Zavaleta, E.S., R.J. Hobbs, and H.A. Mooney. 2001. Viewing invasive species removal in a whole-ecosystem context. Trends in Ecology & Evolution 16:454-459.

Zhang, L., W.R. Dawes, and G.R. Walker. 2001. Response of mean annual evapotranspiration to vegetation changes at catchment scale. Water Resources Research 37(3):701-708.

Zhang, Y. 2006. Development and Testing of a Watershed Forest Management Information System. Version 1.0, Ph.D. Dissertation, University of Massachusetts, Department of Natural Resources Conservation.

Ziegler, A.D., and T.W. Giambelluca. 1997. Importance of rural roads as source areas for runoff in mountainous areas of northern Thailand. Journal of Hydrology 196 (1-4):204-229.

Ziegler, A.D., R.A. Sutherland, and T.W. Giambelluca. 2001. Interstorm surface preparation and sediment detachment by vehicle traffic on unpaved mountain roads. Earth Surf. Process. Landforms 26:235–250.

Ziemer, R.R. 2000. Watershed assessment-watershed analysis: What are the limits and what must be considered. AEG News 43(4):122.

Ziemer, R.R. 1981. Storm flow response to road building and partial cutting in small streams of northern California. Water Resources Research 17(4):907-917.

Appendixes

Appendix A
Institutional Governance and Regulations of Forests and Water

Many reservoirs and other water supply facilities are located on forest lands. Water managers depend upon runoff from upstream forests for water supply and power generation, creating a direct interest in forest management. Unlike land, water is not "owned" outright. Traditional concepts of property ownership and (or) rights do not apply well because water moves across landscapes. Justice Oliver Wendell Holmes wrote, "A river is more than an amenity, it is a treasure. It offers a necessity of life that must be rationed among those who have power over it" [*New Jersey v. New York,* 283 U.S. 336 at 342-44 (1931)]. The legal systems designed to allocate water apply to all forest lands - public and private. They authorize private use, but condition such use on recognition of general public needs, such as navigation, water quality, and recreation. State statutes typically declare that the waters of the state belong to all the people of the state and that water rights are only a right to use water.

WATER RIGHTS AND MANAGEMENT

Three water law systems govern the right to use water in the United States - riparian, prior appropriation, and hybrid rights. Twenty-nine states follow the riparian rule. Nine states follow the prior appropriation system. Ten states follow a hybrid of riparian and prior appropriation law. The remaining two states have unique code rules. These legal systems generally promote diversion and use of water for purposes of meeting human development needs. The basic water allocation systems are run by state water management agencies. Federal, state and local agencies and private organizations must obtain water use rights from the state agencies, except in rare instances.

Federal Water Management

Once water rights are obtained authorizing the use of water, this use is regulated and managed by a wide variety of federal, state and local agencies and private organizations under an equally wide variety of laws. The federal government is deeply involved in water development for multiple purposes including navigation, flood control, hydropower and irrigation.

Federal investment focused on navigation and internal improvements in the early days. The Army Corps of Engineers initially built canals and levees for navigation. The Flood Control Act of 1936 directed the Corps to provide flood protection to the entire country. The Corps maintains and operates 383 dams and reservoirs. Many Corps projects built for navigation or flood damage reduction have additional uses, such as hydroelectric power. The Corps was first authorized to build hydroelectric plants in the 1920s, and today operates 75 power plants, producing one fourth of the country's hydro-electric power, making it the country's fifth largest electric supplier [http://www.usace.army.mil/missions/water.html]. The management and operation of each Corps facility is controlled by the Congressional law that authorized it.

The Bureau of Reclamation was created in 1902 to construct and maintain irrigation facilities to store, divert, and develop water for reclamation of arid and semi-arid lands. With the construction of Hoover Dam and Grand Coulee Dam in the 1930s, the Bureau embarked upon several decades of major project construction (Reisner 1986). The Bureau is currently the largest wholesaler of water in the United States, bringing water to more than 31 million people, and providing one out of five farmers in the western states with irrigation water for 10 million acres of farmland. The Bureau is the second largest producer of hydroelectric power in the western states, with 58 powerplants generating more than 40 billion kilowatt hours, enough electricity to serve 6 million homes. Management and operation of Bureau of Reclamation facilities is controlled by the Reclamation Act itself, additional general statutes and specific statutes authorizing each project.

Many of the large reservoirs in the United States were developed by private companies to generate hydroelectric power. The Federal Power Act of 1920 [16 USC 791-825] allows private development on navigable rivers subject to obtaining a federal license. State and municipal power utilities are also required to obtain federal licenses, typically for 50 years. The licenses for many of these projects are currently expiring. Fish and wildlife and other environmental concerns must be addressed during the renewal process before the Federal Energy Regulatory Commission.

The Army Corps of Engineers, Bureau of Reclamation, and Federal Energy Regulatory Commission must comply with federal environmental laws such as the Endangered Species Act and the Clean Water Act in the operation and maintenance of the array of federal and federally licensed water facilities. Forest managers seeking to achieve fish and wildlife, water supply and watershed protection objectives have an interest in how water facilities are operated. However, management coordination usually occurs only in the context of one agency commenting on another's management plans, participating in consultation under the Endangered Species Act or commenting in other regulatory proceedings. Only under the mandatory consultation and conditioning requirements of the Federal Power Act are land and water managers forced to actually agree upon operating conditions.

Non-Federal Water Management

Some states also own and operate water storage and distribution systems. The most elaborate of these is the California Water Project, the country's largest state-built water and power development and conveyance system. It includes pumping and power plants; reservoirs, lakes, and storage tanks; and canals, tunnels, and pipelines that capture, store, and convey water to 29 water agencies [http://www.water.ca.gov/nav.cfm?topic=State_Water_Project]. State statutes control the purposes, operation and management of these systems similar to how federal laws control the federal projects.

Most municipal and irrigation water is provided by local or regional water service or supply organizations. While some of these providers are private companies, most are municipal corporations or specially designated districts. Private companies generally are regulated by public utility commissions. The municipal corporations often have elected or appointed boards and operate under municipal articles of incorporation and by-laws. These water supply organizations provide water for irrigation and municipal and industrial purposes. Some states authorize larger regional water authorities such as the Northern Colorado Water Conservancy District that provide water to irrigation districts and municipalities.

Water Allocation Systems

Under the riparian doctrine, owners whose land is along rivers and streams (riparian owners) have the right to use water from the waterway in a way that is "reasonable" relative to all other users. If water is insufficient to meet all reasonable needs, all riparians must reduce their use in proportion to their rights. Historically, non-riparians had no right to use water. Most states that follow the riparian doctrine have adopted statutes that require riparian landowners to obtain water permits and allow non-riparian owners to obtain permits as long as their uses do not harm riparian rights (Getches, 1997).

The prior appropriation system developed in the arid states of the west where much of the land was owned by the Federal government at the time of non-native settlement. In order to develop mines, farms and communities, settlers needed to divert water from streams and transport it to where it was needed to support development. The U.S. Congress legally severed water from the land in the 13 Western states [California Oregon Power Co. v. Beaver Portland Cement Co., 295 U.S. 142 (1935)]. Settlers developed a system in which the first person who used water beneficially acquired the right to use the water forever as long as the water was not wasted and was used regularly. Historically, the only uses recognized as beneficial were commercial or consumptive, not uses of water in stream for scenic, recreational, or fish and wildlife purposes. Water rights under the prior appropriation system depend on water usage, not land ownership. Permits are required to appropriate water. Water rights are administered by state agencies (Getches, 1997).

States also administer the permit systems that create water rights and generally enforce the rights between the various public and private water rights holders. When a water right holder seeks to change their water use, generally state approval of the transfer is required. The right to use water in prior appropriation states is forfeited if it is not used. This creates an incentive for water right holders to use as much water as they are authorized to whether they actually need it or not.

Other Considerations

Land and water are inextricably linked, but the laws and institutions governing their use are not. Despite long recognition of the need to manage land use in watersheds across ownership boundaries and across the jurisdictions of multiple layers of government jurisdiction, laws and institutions that truly integrate watershed management have not been created. The fragmentation of ownership and jurisdictions, and the single subject mandates of most agencies, leads to fragmented, conflicting judicial decisions and management.

Integrated watershed management (Chapter 2, Chapter 4) could mitigate the effects of institutional fragmentation; it could also mitigate unintended consequences of cumulative watershed effects. Examples of the problems of fragmentation and cumulative watershed effects (Chapter 4) abound. Notable examples include an Army Corps of Engineer's reservoir drawdown to build a temperature control tower that releases pesticides accumulated in sediments from decades-old U.S. Forest Service spraying; and multiple landowners deciding to harvest timber in a small watershed during the same season, unaware of one another's plans and their cumulative hydrologic effect.

Due to fragmented jurisdictions, governments have been unable to create governance institutions at the watershed level. However, new community-based initiatives and private markets are developing. These place-based voluntary approaches can create a civic space that nurture shared visions for the future of a watershed, implementation of action plans, and tools that knit land and water into an integrated whole for their management.

REGULATION OF FOREST AND WATER USE

Fragmentation of laws and institutions extends beyond forest and water ownership and management to regulation. Separate federal, state and local statutes and ordinances have been adopted to regulate effects of forest and water management. Table 3-1 shows the major regulations applicable to a timber sale on public or private lands, highlighting the complexity and single subject and single agency divisions facing forest land managers.

TABLE 3-1 Water-Related Regulatory Requirements for Timber Harvests

Resource Issue	Non-Industrial Private	Industrial Private	Federal Public Land
Land Use	Local Land Use Plan and Zoning Ordinance	Local Land Use Plan and Zoning Ordinance	Land and Resource Management Plan consistency
Overall impacts	State and local forest management regulations*	State and local forest management and SFI/FSC requirements**	NEPA: Prepare an environmental assessment or an EIS.
Water Quality	Clean Water Act BMPs	Clean Water Act BMPs	Clean Water Act: standards and guidelines from plan
Endangered Species	No take of listed species	No take of listed species	Avoid jeopardizing listed species.
Fire Prevention	State and local regulations	State and local regulations	Standards and guidelines from plan
Wetlands	404 permit	404 permit	404 permit and 401 certification
Herbicides, Insecticides	FIFRA	FIFRA	FIFRA

*SFI or FSC requirements may apply if the product is sold to a mill that requires the standards to be met.
**SFI means Sustainable Forestry Initiative and FSC means Forest Stewardship Council. Both are forest product certification standards.

Early Approaches

Landowners cannot use their property in a way that injures their neighbors' property. Landowners have historically been liable for damages caused by altering the course or amount of water flowing from their land onto adjacent properties and for creating nuisances by unreasonably interfering with their neighbor's use and enjoyment of their property (Mortimer and Visser, 2004).

As concerns grew about the impacts of forest management on hydrology, governments responded by adopting a wide variety of laws and regulations. In 18th and 19th century New England, timber harvesting and log drives destroyed many salmon runs (Montgomery, 2003) and hurt drinking water supplies. States responded by enacting laws to regulate the impact of forest practices by prohibiting or controlling timber harvest in drinking water source areas and dumping of logging wastes in streams, requiring fire control, leaving of seed trees for reforestation and buffers around lakes and along rivers to protect scenery (Cubbage and Siegal, 1985).

Laws regulating forest management now have been enacted at the federal, state and local level. Every forest landowner must comply with those applicable to its forest, requiring managers to follow multiple laws and regulations administered by multiple federal, state and local agencies.

Forest Practice Regulations

The most comprehensive laws regulating forest management have been adopted by the states. The courts have consistently upheld the validity of government regulation of forest management to protect the public interest as long as it does not "take" all use of the property (Cubbage and Siegal, 1985). State forest practices laws were originally enacted in the 1930s and 1940s to require reforestation in order to assure sustained timber production. In the 1960s and 1970s, with increasing concern about environmental issues, several states adopted broader laws aimed at assuring water and air quality, fish and wildlife habitat and scenic quality (Lundmark, 1995).

While these laws vary greatly, they have now been adopted in one form or another in all 50 states. Comprehensive programs administered by a single state agency are referred to as forest practices acts. They are most common in western and northeastern states. Elsewhere it is more common for forest practices to be regulated by a variety of state agencies under multiple different statutes addressing everything from water pollution control to soil erosion to shoreland protection. All of the programs rely upon voluntary or regulatory best management practices (BMPs), sometimes called acceptable management practices, guidelines or forest practice rules.

The programs take three basic approaches: (1) permit inspection systems requiring permits from state agencies before harvest or other operations (e.g., California); (2) notification systems requiring notice to an agency before operations begin and compliance with adopted standards (e.g., Oregon); and (3) contingent systems in which failure to comply with adopted standards results in agency enforcement (Virginia) (Ellefson et al., 1995). A fourth approach requires loggers and/or professional foresters to be licensed or certified. Licensing and certification require training to assure that operators are familiar with BMPs. States with logger certification programs (Connecticut, Maryland, and WestVirginia), often also have enforcement programs with power to revoke a logger's license to practice if BMPs are not followed (Irland and Connors, 1994). Most professional foresters apply BMPs (USEPA, 2005).

State laws and regulations address multiple forest management practices and procedures. Examples of the practices regulated include: reforestation; silvicultural and harvest methods (like restrictions on clearcutting); road construction and maintenance, slash management; fire; chemical use; and forest land conversion. Typical provisions related to effects on hydrology prohibit leaving slash in streams, require riparian buffers and specify road construction methods in riparian areas and on steep slopes. They often prohibit timber harvest in sensitive areas, like wetlands. Some states have explicit rules governing timber harvest and road building in watersheds that provide drinking water.

State forest practice regulations apply on private and public lands. Compliance monitoring is done in most states and surveys show that compliance is generally high. Compliance tends to be lower on private lands than public lands and lower on small private tracts than large industrial ownerships (Ellefson et al., 2001). This raises serious concerns as forest lands are urbanized or acquired by individuals or organizations unfamiliar with forest management. Recent studies conclude that non-industrial private landowners often lack knowledge about BMPs, especially in the southeast where studies show a lack of concern for forest water quality (NCASI, 2006). The Virginia Department of Forestry reports that most landowners make management decisions and sell timber without professional advice, which reduces BMP compliance. In Mississippi, nearly two-thirds of forest lands are owned by non-industrial owners, but a third of the owners were unfamiliar with BMPs (Londo, 2004).

Studies comparing compliance rates under voluntary versus regulatory programs are limited, as are cost comparisons. There is no direct evidence showing that compliance levels are better under mandatory rather than voluntary programs. Evidence does show that costs of both administration and compliance are higher with regulatory programs (Hawks et al., 1993).

Due to federal regulatory developments and policy evolution at the state and local level, states traditionally relying upon voluntary programs have begun to adopt some mandatory regulations. States with comprehensive forest practices acts have also embraced many of the voluntary education and stewardship incentive programs developed by states with voluntary programs. Fundamentally, through regulatory and voluntary programs, state forestry and pollution control agencies have lead responsibility for protecting the public interest in controlling the effects of forest management on watersheds.

National Environmental Policy Act and Cumulative Effects

The National Environmental Policy Act of 1969, 42 U.S.C. 4331 et seq. (NEPA) requires federal agencies to carefully weigh environmental considerations and consider potential alternatives before taking major actions. The heart of NEPA is Section 102 (2)(C) which requires federal agencies to prepare detailed environmental impact statements (EISs) on any major action significantly affecting the quality of the human environment.

Analyzing the environmental impacts of forest management practices on runoff and water quality generates considerable debate at least in part because of legitimate differences of opinion about the probable nature and extent of land-use effects on runoff, biological resources, water quality, and other values. One of the most difficult aspects of environmental impact analysis of forest practices relates to their cumulative effects across ownerships and over time in a watershed.

NEPA requires agencies to consider "whether the action is related to other actions with individually insignificant but cumulatively significant impacts," 40 C.F.R. 1508.27(b). "Cumulative effects" mean:

> "the impact on the environment which results from the incremental impact of the action when added to other past, present, and reasonably foreseeable future actions regardless of what

agency (Federal or non-Federal) or person undertakes such other actions." 40 C.F.R. 1508.7.

Analysis of cumulative effects requires agencies to consider spatial and temporal scales and identification of cause and effect relationships between multiple actions and multiple resources. Two or more forest management actions, like timber harvests, can interact to produce magnified effects on ecosystem functions or other resources, even if each influence alone would have been relatively small or benign (University of California Committee on Cumulative Watershed Effects, 2001). As vital as it is to understanding watershed-scale impacts of forest management, federal forest management agencies have had difficulty doing cumulative effects analysis and their actions have often been challenged (Council on Environmental Quality, 1997).

Timber sale appeals often assert that the agency's Environmental Assessment or EIS is inadequate because the challenged sale is likely to have cumulative effects with other sales planned in a watershed. The USFS has developed Cumulative Watershed Effects models to use in NEPA analysis. Cumulative Watershed Effects are "significant, adverse influences on water quality and biological resources that arise from the way watersheds function, and particularly from the ways that disturbances within a watershed can be transmitted and magnified within channels and riparian habitats downstream of disturbed areas." (University of California Committee, 2001).

The courts have addressed what projects are "reasonably foreseeable," the trigger for requiring cumulative effects analysis. The general rule now is that it is inappropriate to defer cumulative impact analysis to a later date when meaningful consideration can be given, but agencies are not required to do the impractical if not enough information is available to permit meaningful consideration, *Environmental Protection Information Center v. U.S. USFS,* 451 F. 3d 1005, 1014 (9th Cir. 2006).

Although the courts have accepted the USFS's major cumulative watershed effects method, analyzing cumulative effects is difficult and the agency continues to be challenged successfully for inadequate application of its methods. For example, in *Lands Council v. Powell,* 379 F.3d 738 (9th Cir. 2004), the court reversed a decision to implement an aquatic restoration project that involved logging. First it found that the EIS lacked adequate data about the time, type, place, and scale of past timber harvests and should have explained how different project plans and harvest methods affected the environment. A map showing past harvests, with general notes about total acres cut per watershed was not adequate. Second, the court found it arbitrary for the USFS to rely on a particular instream sedimentation model (the Water and Sediment Yields ("WATSED") model) that it knew had limitations, without disclosing the limitations.

Land-use signals may be hard to define in quantitative terms. Cumulative effects continue to be of great concern to resource managers and regulators in forested mountain regions, where the goals of timber harvest may conflict with other social goals for water quality or biodiversity (University of California Committee, 2001).

From the early days of NEPA implementation, the Council on Environmental

Quality and others believed that NEPA analysis would be a way to assure integrated analysis of the effects of forest management actions on the environment. Earlier guidance urged agencies to integrate the NEPA process into other planning at the earliest possible time to ensure that planning and decisions reflect environmental values, to avoid delays later in the process, and to head off potential conflicts (CEQ, 1981). Despite widespread desire to integrate multiple environmental requirements, especially those under the Clean Water Act and the Endangered Species Act, with NEPA, no generalized method has been developed to do so. Instead agencies rely on case-by-case and agency-by agency coordination with varying decrees of success (CEQ, 2003).

Clean Water Act

Public concern about the effects of forest management on hydrology has focused primarily on water quality. Congress attempted to control dumping into rivers and streams as early as 1899. This early law prohibited dumping logging slash and debris in navigable waters. But no comprehensive system of water pollution control existed until 1948 with the Federal Water Pollution Control Act. This Act was substantially amended in 1972 and 1987 and is now known as the Clean Water Act (CWA, 33 USCA, Section 1151 et seq., 1972). Its stated objective is to restore and maintain the chemical, physical, and biological integrity of the nation's waters. It sets a national goal to attain water quality that "provides for the protection and propagation of fish, shellfish, and wildlife and provides for recreation in and on the water." 33 USC § 1251(a). The Environmental Protection Agency (EPA) has overall responsibility for implementing the CWA. Congress intended the Act to be implemented by the states, and states may adopt stricter regulations.

The CWA requires states to adopt water quality standards for all water bodies including rivers, streams, wetlands, lakes and reservoirs. The standards are designed to assure that water is clean enough to allow specific designated beneficial uses, such as human consumption or fish spawning. Numeric or narrative criteria are then adopted for various pollutants to protect the most sensitive of the designated beneficial uses (usually fish spawning and rearing). Existing high quality water must be protected from degradation. New sources of pollution are prohibited in waters that already fail to meet the standards.

Section 313 requires that federal activities must meet state water pollution control requirements. Thus, timber harvest and other activities on federal forest lands must meet state water quality standards (*Northwest Indian Cemetery Protective Association v. Peterson,* 565 F Supp 586, *aff'd in part, vacated in part,* 764 F2d 581 (9th Cir. 1985)). The district court found that a proposed road and timber harvest on a National Forest would violate water quality (see Chapter 3) standards for turbidity and sediment and adversely affect the designated beneficial use of habitat and spawning for anadromous fish.

The major method Congress created to control water pollution was to require point sources of pollution to obtain discharge permits, known as National Pollutant

Discharge Elimination System (NPDES) permits, before adding pollutants to waters of the United States. Generally, any discrete conveyance, such as a pipe or a ditch, is a point source. NPDES permits include permit conditions and prohibit discharges that would violate water quality standards. Most industries, wastewater treatment plants and stormwater systems now have NPDES permits.

Non-point Source Pollution

EPA regulations exclude from the NPDES permit requirement "non-point source silvicultural activities." (40 CFR 122.27(b)(1)). Silvicultural point sources requiring NPDES permits are limited to discernible, confined and discrete conveyances related to rock crushing, gravel washing, log sorting, or log storage facilities. Nonpoint source silvicultural activities do not require permits. They include activities like site preparation, reforestation, thinning, prescribed burning, pest and fire control, harvesting operations, surface drainage, or road construction and maintenance from which there is natural runoff. Federal courts are now split on whether stormwater systems on forest roads require permits.

Congress expected water quality impacts of silvicultural activities to be addressed by state programs aimed at controlling nonpoint sources of pollution. Under the 1972 FWPCA states developed areawide water-quality management plans under Section 208. All states were to assess damages to water quality from nonpoint source pollution and to develop either regulatory or non-regulatory programs to control them. EPA turned to Best Management Practices (BMPs) as the way to address the water quality impacts of forest management. States adopted forestry BMPs in their 208 plans as a way to meet the goals of their water quality management plans, especially after passage of the Federal Water Pollution Control Act in 1977 (Ice et al., 1998).

EPA defines Best Management Practices to be the methods, measures and practices that prevent or reduce non-point source pollution. EPA recommends that to the extent possible, best management practices be implemented before, during and after forest management activities to reduce or eliminate the introduction of pollutants. BMPs generally are based on a substantial body of research; however, BMPs are incomplete or unknown for some forest management practices and/or in some regions. Most BMPs (required or voluntary) focus on minimizing sediment or temperature increases and dissolved oxygen from harvesting operations. Current BMPs cover a range of water quality effects and are routinely applied in categories similar to those under forest practices acts: streamside management zones with limited harvesting; clearcut size limits; site preparation and reforestation requirements; road building and skid trail location restrictions; high hazard site/steep slope restrictions to limit erosion and landslides; wetlands protections; and regulation of fertilizer and pesticide application (USEPA, 2005). The issues facing managers today relate to the adequacy of existing BMPs, particularly the need to measure and quantify results of BMP implementation.

Total Maximum Daily Load Allocations

Section 303(d) requires states to identify waterbodies that fail to meet water quality standards. A "total maximum daily load" (TMDL) must be developed for each of the impaired waters. 33 U.S.C. 1313(d). A TMDL determines how much assimilative capacity exists in a water body and then allocates portions of that capacity to point sources, non-point sources and a safety factor for future growth. Point source allocations are implemented through amendments to NPDES permits. Nonpoint source allocations are implemented through the Section 319 Programs.

If a state fails to prepare TMDLs, EPA must prepare a priority list for the state and develop its own TMDL determination. Most states lacked the resources to do TMDL analyses, which involve complex assessment of point and nonpoint sources and mathematical modeling. The major controversy surrounding TMDLs is whether states can, or must, impose specific, quantified load allocations on nonpoint sources, like silviculture. The USFS protested application of TMDLs to forest management, arguing that forest practices would expose the agency to litigation over nonpoint source controls, and that BMPs, without specific limits on pollutants, were more appropriate (Houck, 2002). The National Association of State Foresters and the Society of American Foresters share this position. EPA has adopted rules, however, that require states to address nonpoint sources of pollution in TMDLs.

The result of the TMDL program and these judicial rulings, however, has been that as point source dischargers face expensive compliance requirements in order to meet water quality standards under TMDLs, they advocate stronger state regulation of nonpoint sources, like silviculture, and seek mechanisms to meet their TMDL requirements by investing in nonpoint source controls if they are less expensive than further reductions in the point source discharges. This has prompted development of environmental service markets, as discussed below.

Section 404 Wetland Regulations

Wetlands are recognized by the CWA as an important and dwindling resource. A significant portion of freshwater wetlands loss occurs on forested wetlands (Dahl and Johnson, 1991). Forested wetlands are difficult to create or restore if they are lost (NRC, 2001). Federal policy is to prevent destruction of wetlands for anything but water-dependent activities or aquaculture. Anyone proposing to fill wetlands must obtain a permit from the U.S. Army Corps of Engineers under Section 404 of the CWA; EPA has review and policy-setting functions over both the Corps and the states (Want, 2006).

"Waters of the United States" has been expansively interpreted and includes rivers, streams, lakes, estuaries and swamps or wetlands. In 2001, however, the Supreme Court held that "isolated wetlands" which are not part of, or adjacent to surface tributary systems, are not included. Even if federal jurisdiction extends to a particular wetland, Congress exempted "normal farming, silviculture and ranching activities" from the permit requirement, 33 U.S.C. 1344(f)(1). The exemption does not apply, however, if the proposed activity is to change the land use and the reach of the water is reduced due to the change, 33 U.S.C. 1344(f)(2). The Corps and the EPA also interpret the exemption to require that silviculture be part of an "established, ongoing operation," not for the conversion of wetlands to forest management (EPA, 1990 and 40 C.F.R. 232.3(b)). A separate exemption exists for constructing and maintaining forest roads and for activities regulated by approved best management practices under Section 208.

These exemptions are important for forest management, especially in the south and in coastal areas, since conversion of farmland to pine plantations is a common practice. Since the land use is changed, the exemption does not apply. Removing vegetation to clear land requires a 404 permit if land levelling or substantial earth-moving occurs in wetlands. If earth is moved to replace an aquatic area with dry land, it is considered to be fill, and a permit must be obtained first. The Corps and EPA have issued special guidance on silvicultural site preparation activities regulating when formerly agricultural lands are converted to pine plantations. Permits are required for mechanical silvicultural site preparation for nine types of wetland areas, such as riverine bottomland hardwood wetlands, white cedar swamps and swamp forests (EPA, 1995), however, if BMPs are followed, no 404 permit is required. If the wetlands have been so altered through past practices that they no longer function as wetlands, no permit is required either. The Reauthorization Amendments of 1990 (16 U.S.C. §§ 1455(d)(16), 1455b). It required coastal states to adopt enforceable mechanisms to control activities causing or contributing to nonpoint source pollution in the coastal zone.

Safe Drinking Water Act

As early as 1808 laws protected drinking water sources by regulating activities in watersheds or barring human entry into them. Cities sought pure drinking water sources, often on forested lands, where they built reservoirs to supply their citizens. In the 19th Century, protection of water purity at the source was accepted practice. The ideal water source was one "free from human habitation and is covered with forest." As treatment technology improved, however, water suppliers shifted from this watershed protection approach to reliance on treatment, "purified water" rather than "pure water." (Porter, 2006).

In 1974, Congress enacted a new statute to protect drinking water in response to outbreaks of waterborne disease and increasing chemical contamination of public water sources and concerns about aesthetics and taste. The Safe Drinking Water Act (SDWA) authorizes the EPA to set maximum contaminant levels (MCLs) in public drinking water. In 1989, EPA adopted rules requiring surface water systems to filter their water, unless they can prove that filtration is unnecessary. Unfiltered systems must have an effective watershed control program. In 1996, recognizing that treatment alone was not addressing all problems and often was extremely expensive, Congress added requirements that water suppliers prepare Source Water

Assessments to tell consumers where their water comes from, what contaminants are in it, and whether the water poses a risk to health. The renewed emphasis on protecting drinking water at the watershed level reflects public concern that traditional water treatment methods may not be adequate to assure public safety (Porter, 2006). Source Water Assessments should identify risks to all water resources used (or to be used) as drinking water supplies. Every state has now developed a Source Water Assessment Plan that sets priorities and lays out a process for completion of the assessments. EPA lists several forest practices as potential sources of contamination including harvesting, residue management, fertilization, pesticide application and road construction and maintenance (EPA, 2005).

Other State and Federal Regulations

Many additional federal and state laws and regulations can apply to any given forest management action. Each one is administered by a separate agency, or division within an agency, dedicated to protecting the specific public interest the statute addresses. For example, the Endangered Species Act (ESA) (16 U.S.C. §§ 1531-1544) administered by the U.S. Fish and Wildlife Service (USFWS) and the National Marine Fisheries Service (NMFS) has become a significant factor in forest and water management on public and private lands because of the importance of aquatic and riparian habitat to many listed species. As a result, timber harvesting, livestock grazing, road construction and many other forest management activities have been curtailed or modified due to ESA requirements. Another example is the Federal Insecticide, Fungicide and Rodenticide Act (FIFRA), which is administered by EPA or states with delegated authority regulates application of chemicals to forests for purposes such as insect and disease control and weed control during reforestation to assure that the pesticides do not remain in the soil, air or water in quantities that could harm water quality or fish and wildlife. There are also state and federal laws that protect scenery along rivers and lakes, often restricting forest harvesting within buffer zones. For example, the federal Wild and Scenic Rivers Act, 16 USC 1271-1287, protects designated free-flowing rivers that have "outstandingly remarkable scenic, recreational, geologic, fish and wildlife, historic, cultural and other similar values."

ENVIRONMENTAL SERVICES MARKETS

Existing institutions and legal systems have not kept pace with public recognition of the benefits provided by forests. These values are now recognized as "environmental services." (Pagiola et al., 2002). They include watershed protection, biodiversity conservation and carbon sequestration (Jenkins et al., 2004). Forests' watershed protection services, such as water quality, flow regulation, water supply, flood prevention, salinization control and aquatic habitat, are among the most valuable. For example, cities which depend upon unfiltered water estimate that every

$1 invested in watershed protection can save anywhere from $7.50 to nearly $200 in treatment and filtration costs (Reid, 1997).

A recent international study identified 61 efforts to establish markets for the watershed services forests provide. The study illustrates how local communities, private companies and individual landholders are creating new market mechanisms to deliver and finance watershed protection. These local and regional efforts attempt to "ensure that land managers internalize the negative impacts they have on water quality and flow." Failure of traditional governmental approaches has lead to local innovation (Landell-Mills and Porras, 2002). In all environmental service market situations there is a need to better quantify the results of various treatments and apply simulation models to relate treatment effects to downstream areas or "markets" in order to begin to use environmental service markets

Environmental services markets have developed for many reasons, but fundamentally they are a response to the failure of markets to value the services forests provide and the high cost of traditional governmental approaches to forest conservation. Agencies lack funds to acquire forest lands in order to protect their watershed functions. Frustration has grown with the regulatory approaches discussed above because they are often inefficient, expensive and inequitable. Private landowners interested in managing their lands to provide clean water and stable water flows have no incentive to do so. The beneficiaries of the "services" provided forests are usually downstream and have no reason to pay for services that have traditionally been free.

The market mechanisms for forest landowners to manage for water benefits can be categorized into three types: (1) self-organized private deals; (2) open trading systems; and (3) public payment systems (Powell et al., 2002). Private deals include all direct transactions between beneficiaries of forest management and forest landowners who provide them. Purchase of conservation easements and development rights are the oldest and most pervasive private forest conservation transaction. Examples include water utilities buying land to protect drinking water supplies or land trusts buying property to protect wetlands or other watershed functions.

Open trading markets are the second category of environmental services markets, such as the markets that have developed for carbon offsets since adoption of the Kyoto Protocol. The first step in creating such markets is to quantify the amount of environmental service provided by a particular forest management practice. What service is demanded and how can it be measured? What management practice creates the "product" or service? Once the service is "commodified," in this way, public or private markets between buyers and sellers of the service can develop. The most prevalent markets for watershed services provided by forests are wetland mitigation banks and effluent trading systems.

Water Quality Trading

Water quality trading develops when one party, usually an industrial facility or wastewater treatment plant, faces relatively high pollutant reduction costs. If pollutant loads in the receiving water can be reduced for less by investing in forest restoration or other forest practices, it is worthwhile for the point source to compensate a forest landowner to achieve the less costly pollutant reduction by planting riparian buffers or reforesting lands. Several water quality trading markets are currently operating, with others under development. Most of these markets are focused on either phosphorus or nitrogen-based trading, such as the Tar-Pamlico Basin in North Carolina and the Lower Boise River in Idaho. Trading is also underway for various forms of salmon habitat, salinity, temperature and transpiration (Landall-Mills and Porras, 2002).

Mitigation Banks

Mitigation banking means the "restoration, creation, enhancement and, in exceptional circumstances, preservation of wetlands and/or other aquatic resources expressly for the purpose of providing compensatory mitigation in advance of authorized impacts to similar resources." (Federal Register, 1995) The objective of a mitigation bank is to replace the functions of wetlands and other aquatic resources which are lost due to filling or other activities authorized under Section 404 or other permits. When the "bank" is established, the functions it provides are quantified as mitigation ``credits'' which are then available for use by the bank sponsor or by other parties to compensate for adverse impacts (i.e., ``debits''). Credits may only be used by permittees when impacts to aquatic resources are unavoidable and on-site compensation is either not practicable or use of a mitigation bank is environmentally preferable to on-site compensation.

A developer seeking to fill wetlands, for example, could buy "credits" from a bank which had been established previously by restoring other wetlands. This often allows significant wetlands restoration to be funded by the cumulative credit purchases of many small developments, which can offer significant efficiency and ecological benefit. By 2000 over 70 commercial wetland mitigation banks were operating in the United States. Costs of the credits range from $7500 per acre to as much as $100,000 per acre (NRC, 2001).

The third category of markets involves direct public payments to landowners for environmental services. Examples of this type include the conservation reserve and wetland reserve programs under the Farm Bill. Under this approach, the government pays landowners directly to set aside and manage their lands to reduce erosion and runoff.

Many unresolved questions exist in the development of markets for the watershed services provided by forests. Key questions remain about the exact nature and value of the service provided. For example, it is very difficult to quantify to the level of temperature reduction provided by a riparian buffer, or how much that reduction is worth. When markets develop there must also be ways for buyers, sellers and regulators to measure and monitor the services provided.

CLOSING

This appendix provides a brief summary of how water resources, including those that are outputs from forests, are governed in the United States. It traces the origins of the primary pieces of legislation that shape how water is managed and governed; it also elucidates how the governance of water and forests was and has remained fragmented, despite the physical interconnectedness of forests and water. This appendix presents a rationale for how forests may be included in some of these laws and regulations and offers ways to consider forests, and water as an elemental forest output, as potential components in environmental service markets.

REFERENCES

Council on Environmental Quality. 1997. Considering Cumulative Effects under the National Environmental Policy Act. Washington, D.C.

Cubbage, F.W., and W.C. Siegel. 1985. The Law Regulating Private Forest Practices. Journal of Forestry. __:538-545.

Dahl, T. E., and C.E. Johnson, Status and Trends of Wetlands in the Coterminous United States, Mid-1970s to Mid-1980s (1991), U.S. Department of the Interior, Fish and Wildlife Service.

Ellefson, P.V., A.C. Cheng, and R.J. Moulton. 1995. Regulation of Private Forestry Practices by State Governments. Minnesota Agricultural Experiment Station (Bulletin 605-1995).

Ellefson, P.V., M.A. Kilgore, and M.J. Phillips. 2001. Monitoring Compliance with BMPs: The Experience of State Forestry Agencies. Journal of Forestry. 11-17.

Forest Environmental and Sustainability Issues in the Southern United States. Special Report No. 06-06. NCASE Southern Regional Meeting. Ashville, North Carolina.

Getches, D.H. 1997. Water Law in a Nutshell. Third Edition. St. Paul, MN: West Publishing Company.

Hawks, L.J., F.W. Cubbage, H.L. Haney, Jr., R.M. Shaffer, and D.H. Newman. 1995. Forest Water Quality Protection: A Comparison of Regulatory and Voluntary Programs. 91:48-54 J. Forestry.

Houck, O. 2002. The Clean Water Act TMDL Program: Law, Policy, and Implementation (2d ed.). Environmental Law Institute. Washington, D.C.

Ice, G.G., G.W. Stuart, J.B. Waide, L.C. Irland, and P.V. Ellefson. 1998. 25 Years of the Clean Water Act: How Clean Are Forest Practices? J. Forestry 9-13.

Irland, L.C., and J.F. Connors. 1994. State Nonpoint Source Programs Affecting Forestry: The 12 Northeastern States. Northern Journal of Applied Forestry. 11: 5-

Jenkins, M., S.J. Scherr, and M. Inbar. 2004. Markets for Biodiversity Services:

Potential Roles and Challenges. Environment. 46:32-42.

Landell-Mills, N., and I.T. Porras. 2002. Silver Bullet or Fool's Gold? A Global Review of Markets for Forest Environmental Services and Their Impact on the Poor. London: International Institute for Environment and Development.

Lundmark, T. 1995. Methods of Forest Law-Making, Boston College Environmental Affairs Law Review. 22:783.

National Council for Air and Stream Improvement. 2006. A Primer on the Top Ten.

NRC (National Research Council). 2001. Compensating for Wetland Losses under the Clean Water Act. National Academy Press, Washington, D.C.

Pagiola, S., J. Bishop, and N. Landell-Mills, eds. 2002. Selling Forest Environmental Services. Earthscan. London.

Porter, K. 2006. Fixing Our Drinking Water: From Field and Forest to Faucet. Pace Environmental Law Review. 23: 389-422.

Powell, I., A. White, and N. Landell-Mills. 2002. Developing Markets for the Ecosystem Services of Forests. Forest Trends. Washington, D.C.

Reid, W.V. 2001. Capturing the value of ecosystem services to protect biodiversity. In Managing human-dominated ecosystems, eds. G. Chichilenisky, G.C. Daily, P. Ehrlich, G. Heal, J.S. Miller. St. Louis: Botanical Garden Press.

Reisner, M. 1986. Cadillac Desert. New York: Viking.

USEPA (U.S. Environmental Protection Agency). 2005. National Management Measures to Control Non-point Source Pollution from Forestry. (EPA-841-B-05-001, May 2005).

USEPA. 1995. Memorandum to the Field from the Corps and EPA Regulatory Program Chiefs, Subject: Application of Best Management Practices to Mechanical Silvicultural Site Preparation Activities for the Establishment of Pine Plantations in the Southeast (Nov. 28, 1995). Available online at: http://www.epa.gov/owow/wetlands/silv.html.

University of California Committee on a Scientific Basis for the Prediction of Cumulative Watershed Effects. 2001. A Scientific Basis for the Prediction of Cumulative Watershed Effects. University of California Wildland Resource Center Report. No. 46.

Want, W.L. 2006. Law of Wetlands Regulation (Environmental Law Series). Thomson/West.

Appendix B
Committee Biographical Information

PAUL K. BARTEN, *Chair*, is Associate Professor of Forest Resources at the University of Massachusetts Amherst. He also serves as Co-Director of the USDA Forest Service-University of Massachusetts Watershed Exchange and Technology Partnership. He was a Bullard Fellow in Forest Research at the Harvard Forest (2003-04) while on sabbatical leave. He was a faculty member in the School of Forestry and Environmental Studies at Yale University from 1988 to 1997. Dr. Barten earned a Ph.D. (1988) and M.S. (1985) in forest hydrology and watershed management at the University of Minnesota; he has undergraduate degrees in forestry from the SUNY College of Environmental Science and Forestry (B.S. 1983) and the New York State Ranger School at Wanakena (A.A.S. 1977). Dr. Barten teaches an undergraduate course in forest conservation and graduate courses in forest and wetland hydrology and forest resources management. His research and outreach work focuses on forest hydrology, watershed modeling, and watershed management in the northeastern United States. He was a member of the NRC Committee to Review New York City's Watershed Management Strategy (1997-2000) and the NRC Committee on Conservation of Atlantic Salmon in Maine (2001-04). He also serves on the Research Planning Committee of the Sustainable Forest Management Network in Canada, as chairman of the Massachusetts Forestry Committee, and as a member of the Board of Trustees of the Great Mountain Forest in northwestern Connecticut.

JULIA A. JONES, *Vice-Chair*, is professor at the Department of Geosciences, Oregon State University. Prior to coming to OSU in 1991 she was tenured faculty at the University of California, Santa Barbara. Her research addresses human land use and natural disturbance effects on fluxes of water, sediment, wood, and exotic plant propagules in managed forest landscapes, using long-term records and novel analytic approaches. Dr. Jones the director of the Ecosystem Informatics IGERT and strategic initiative at Oregon State University, co-investigator on the H.J. Andrews Long-Term Ecological Research program, and Director of the Geography Program at Oregon State University. She received a B.A. in economic development from Hampshire College; an M.A. in international relations from the Johns Hopkins School for Advanced International Studies, and a Ph.D. in geography/environmental engineering from The Johns Hopkins University.

GAIL L. ACHTERMAN directs the Institute for Natural Resources at Oregon State University. Her research interest focuses on use of scientific information in public policy making, collaborative decision processes and community water planning. She is an adjunct professor of forest resources and a member of the graduate water faculty. Prior to coming to the university, she worked as an attorney in the Solicitor's Office of the U.S. Department of the Interior in Washington, D.C. where she advised the Bureau of Land Management and the Bureau of Reclamation from 1975-1978. Upon returning to Oregon she joined what later became the Stoel Rives law firm in Portland, where she helped build the Northwest's first specialty natural resource and environmental law practice. She also started her civic work by serving on the Oregon Water Policy Review Board from 1981-1985. In 1987, she became assistant to the governor for natural resources. In that position, she was responsible for natural resource, energy and environmental policy development and implementation, and worked closely with all state resource agencies. She also served as the main liaison with federal resource agencies, including extensive work on plans for Oregon's national forests. She returned to private law practice at Stoel Rives until 2000 when she became Executive Director of the Deschutes Resources Conservancy in Bend, Oregon, working on developing voluntary, market-based watershed restoration methods. She received her A.B. in economics with honors from Stanford University; an M.S. in natural resources policy and management, University of Michigan and a J.D. cum laude from the University of Michigan.

KENNETH N. BROOKS is professor of forest hydrology and watershed management in the Department of Forest Resources, College of Food, Agricultural and Natural Resource Sciences, University of Minnesota. Following five years with the U.S. Army Corps of Engineers, including two years with the Hydrologic Engineering Center, he joined the faculty of the Department of Forest Resources in 1975. During his tenure with the University, he has taught courses in forest and wetland hydrology, watershed management, agroforestry, and range management. His research has focused on forest hydrology, wetland hydrology, hydrologic modeling, and methods of evaluating/appraising watershed management practices. He is a professional hydrologist, certified by the American Institute of Hydrology (AIH), and has served as the Chair of the Board of Registration of AIH from 1997-2003. He was a Fullbright Lecturer from 1997-1998 in Taiwan, and served as a Fellow, East-West Center, Honolulu, Hawaii, from 1984-1985. He received a B.S. in range science-watershed management from Utah State University and an M.S. and Ph.D. in watershed management from the University of Arizona.

IRENA F. CREED is a professor in the Department of Biology and in the Graduate Program in Environmental Sciences at the University of Western Ontario. She is the founder and leader of the Catchment Research Facility (CRF), an advanced monitoring, analytical and modeling facility established for the analysis of catchment processes. Her research interests are in the areas of ecology, hydrology, biogeochemistry, geographic information systems, remote sensing, and simulation modeling. Specifically, Dr. Creed investigates the dominant factors regulating energy, water, and nutrient processes and pathways within specific watersheds in a range of forest regions and biomes.

PETER F. FFOLLIOTT is professor at the School of Natural Resources and Arid Lands Resource Studies, College of Agriculture and Life Sciences, and the Laboratory for Tree-Ring Research, College of Science, University of Arizona. His research involves areas such as watershed, forest, and range and wildlife management, agroforestry, mensuration and inventory techniques, economic assessments and evaluations, ecosystem modeling and simulation, environmental and natural resources policies. Prior to coming to the University of Arizona in 1970, he was a Research Forester with the USDA Forest Service, Flagstaff, Arizona. Dr. Ffolliott teaches and conducts research programs in support of water resources, watershed, and other natural resources management. He is a Fellow of the Society of American Foresters, the Indian Association of Hydrologists, and the Arizona-Nevada Academy of Science, and a member of other many other professional and honorary societies. He received a B.S. and M.S. in forest management from the University of Minnesota; and a Ph.D. in watershed management and water resources administration from the University of Arizona.

ANNE HAIRSTON-STRANG has been a forest hydrologist with the Maryland Department of Natural Resources Forest Service in Annapolis, MD since 1997. She heads the Forest Watershed Management Program, which includes watershed forest restoration and streamside buffer establishment. Her current projects include forest management plan development for Washington Suburban Sanitary Commission forest lands around two drinking water reservoirs, riparian forest buffer monitoring, harvesting BMP assessment, and state coordination with the Chesapeake Bay Program Forestry Work Group. Hairston-Strang received a B.S. in forest management from Virginia Tech; an M.S. in forest soils from University of Minnesota, and a Ph.D. from Oregon State University in forest hydrology.

MICHAEL C. KAVANAUGH is a chemical and environmental engineer, consulting professor at Stanford University, and vice president of Malcolm Pirnie, Inc. His research interests are hazardous waste management, soil and groundwater remediation, process engineering, industrial waste treatment, technology evaluations, strategic environmental management, compliance and due diligence auditing, water quality, water and wastewater treatment, and water reuse. He has served as chair to the National Research Council's (NRC) Board on Radioactive Waste Management and the Water Science and Technology Board. He has also chaired the NRC Committee on Ground Water Cleanup Alternatives. Dr. Kavanaugh is a registered chemical engineer in California and Utah, a Diplomat (DEE) of the American Academy of Environmental Engineers, and a member of the National Academy of Engineering. He received his B.S. in chemical engineering from Stanford University and his M.S. and Ph.D. from the University of California, Berkeley.

LEE MACDONALD is a professor in the Department of Forest, Rangeland, and Watershed Stewardship at Colorado State University. His general field of interest is land use hydrology, broadly defined as the effects of changes in land use on the quantity, timing and quality of runoff. His earlier work focused on hillslope hydrology and the effects of forest management on runoff, and more recently he has been emphasizing the effects of unpaved forest roads and fires on erosion. His process-based work at the hillslope scale has been coupled with studies on hillslope stream-connectivity, and the extent to which these upstream changes can induce downstream cumulative effects. He has a continuing interest in watershed-scale monitoring, our ability to detect significant change, and how this may affect our ability to apply the principles of adaptive management at the watershed scale. His degrees include a B.A. in human biology from Stanford University; an M.S. in resource ecology from the University of Michigan, Ann Arbor; and a Ph.D. in wildland resource science from the University of California at Berkeley.

RONALD C. SMITH is the managing forester at Tuskegee University, where he is a teaching faculty member; oversees the management of the University's land holdings; and serves as Director for Forestry and Natural Resources at the Small Farm Rural Economic Development Center. In his faculty position in the College of Agricultural, Environmental and Natural Sciences, he teaches dendrology (tree identification and silvics) and industrial forestry in the Department of Agriculture and Natural Sciences, and runs the extension forestry and farming programs through the University. He also works with students and other foresters through the Alabama Consortium of Forestry Education and Research on a range of research projects that include valuation of non-timber benefits of different forest ecosystem management strategies, the impact of site preparation on timber and non-timber values, wetland function and ecosystem services, and other topics related to forested systems. Smith was a forester for the USFS from 1979 to 2001 and spent many of those USFS years as an adjunct faculty member at Tuskegee teaching students the trade of forestry. He received his degree in forest management science.

DANIEL B. TINKER is an assistant professor at the University of Wyoming. His research is conducted in the Greater Yellowstone-Teton Ecosystem in northwestern Wyoming, and involves ecosystem responses to large, natural disturbances such as fire. Dr. Tinker uses GIS and remote sensing to investigate the consequences of landscape-scale spatial heterogeneity in ecological systems. His current work is focused on understanding how the observed variation in post-fire plant communities in the Greater Yellowstone-Teton Ecosystem affects important ecosystem processes such as decomposition and nitrogen mineralization, how these processes vary at the landscape scale, and how the effects of post-fire community structure change over time in young, developing forests. He received his B.S. from Fort Lewis College, and M.S. and Ph.D. from University of Wyoming.

SUZANNE BIRMINGHAM WALKER is Vice President and forester/biologist of Azimuth Forestry Services, Inc., manages non-industrial private forestland and provides biological consulting for private land holders in east Texas and western Louisiana. She also produces biological evaluations and environmental assessments for government agencies and privately held companies, wetland delineations, botanical surveys under cooperative agreements, and species checklists of the flora of the Big Thicket National Preserve with Rice University. Prior to Azimuth, she was in the timber industry as a technical cartographer and as a procurement forester in the 1980s, and spent eight years as a federal forester and ecologist with the U. S. Forest Service, where she conducted environmental assessments, timber prescriptions, and biological evaluations for threatened and endangered species and management of wildland fire. She has served as member for the State of Texas Prescribed Burning Board since 1999. Ms. Walker received her bachelors in forestry and biology from Stephen F. Austin State University.

BEVERLEY C. WEMPLE is an associate professor of Geography and Geology at the University of Vermont. Her research interests lie in understanding the dynamics of hydrologic and geomorphic processes in upland, forested watersheds. She is particularly interested in using basic theoretical tools and simulation modeling, in conjunction with empirical field studies, to understand how management of forested, mountain landscapes alters the processes of runoff generation and sediment production. Dr. Wemple is a cooperator with the Long-term Ecological Research network, the Northeastern Ecosystem Research Cooperative, and the Vermont Monitoring Cooperative. She received her B.A. from the University of Richmond, and her M.S. and Ph.D. from Oregon State University.

GEORGE H. WEYERHAEUSER, JR. is elected senior vice president of technology in Weyerhaeuser Company. Since 1978, he has held various positions, including technical forester, contract logger administrator, sawmill supervisor, and vice president and mill manager for containerboard. In 1990 Weyerhaeuser moved to the company's headquarters to become Vice President, Manufacturing for Weyerhaeuser cellulose fibers and paper businesses. He served as president and chief executive officer of Weyerhaeuser Canada Ltd. from 1993-1998. Weyerhaeuser currently serves as a director of Clearwater Management Company and president of the Thea Foss Waterway Public Development Authority. He is also a board member of the Institute of Paper & Science Technology (IPST) and the Museum of Glass: International Center for Contemporary Art. He received his B.A. in philosophy/mathematics from Yale University and his M.S. from Massachusetts Institute of Technology.

management of forested, mountain landscapes alters the processes of runoff generation and sediment production. Dr. Wemple is a cooperator with the Long-term Ecological Research network, the Northeastern Ecosystem Research Cooperative, and the Vermont Monitoring Cooperative. She received her B.A. from the University of Richmond, and her M.S. and Ph.D. from Oregon State University.

GEORGE H. WEYERHAEUSER, JR. is elected senior vice president of technology in Weyerhaeuser Company. Since 1978, he has held various positions, including technical forester, contract logger administrator, sawmill supervisor, and vice president and mill manager for containerboard. In 1990 Weyerhaeuser moved to the company's headquarters to become Vice President, Manufacturing for Weyerhaeuser cellulose fibers and paper businesses. He served as president and chief executive officer of Weyerhaeuser Canada Ltd. from 1993-1998. Weyerhaeuser currently serves as a director of Clearwater Management Company and president of the Thea Foss Waterway Public Development Authority. He is also a board member of the Institute of Paper & Science Technology (IPST) and the Museum of Glass: International Center for Contemporary Art. He received his B.A. in philosophy/mathematics from Yale University and his M.S. from Massachusetts Institute of Technology.